Sports and Traumatology

Series editor

Philippe Landreau, Qatar Orthopaedic and Sports Medicine Hospital, Doha, Qatar

More information about this series at http://www.springer.com/series/8671

Sébastien Gnecchi • François Moutet

Hand and Finger Injuries in Rock Climbers

Springer

Sébastien Gnecchi
Marseille
France

François Moutet
Grenoble
France

Translated by Lucette Cayol
Original French edition published by Springer, Paris, 2010.

ISSN 2105-0759 ISSN 2105-0538 (electronic)
Sports and Traumatology

ISBN 978-3-319-16789-3 ISBN 978-3-319-16790-9 (eBook)
DOI 10.1007/978-3-319-16790-9

Library of Congress Control Number: 2015939680

Springer Cham Heidelberg New York Dordrecht London

Printed on acid-free paper

Springer International Publishing AG Switzerland is part of Springer Science+Business Media
(www.springer.com)

Collection Sports and Traumas Supervised by Philippe Landreau

People, whatever their age, keep practicing sports, and consequently, injuries resulting from sports practice have become more and more specific, which accounts for a new field to be highly considered. Indeed, each joint or part of the body can suffer from any kind of injury. Moreover, each sport has its own specific lesions.

This is why *Sports and Injury Collection* insists on developing each topic according to the sport practiced. This documentation is addressed to all readers, doctors, surgeons, or physicians looking for information about injuries and treatments.

This collection is co-published by the French Society of Sports Traumas (FSST) and the European Federation of National Associations of Orthopaedic sports Traumatology (EFOST).

Prologue

Photo 1 Jacky Godoffe climb to Fontainebleau

Nimble fingers eyes like a hawk

As a child, fingers are first used "to pick one's nose" after the entire house is hand painted with some leftover found at the back of the garage. And next, they are gradually easier to handle once you realize how useful they can be.

Especially in rock climbing, since, as Edlinger put it, "you can't climb but bare-handed."

From that moment, it's just like a musician who couldn't live without his music; the hand becomes indispensable; a failure in one of the fingers, and everything would go wrong; no more sensation, no more pleasure. In a word, especially for a passionate person, the world would be falling apart, and that really sucks!

I've been rock climbing for more than 20 years now, so what should I tell you to take care of yourself?

Perhaps should I mention a slight straining at "la Rose et le Vampire" in Buoux 15 years ago. I shouldn't have tried a hard route after going through the pressure of a former competition because even if I won it, my body eventually failed me.

It served me right. For 3 months, I went on climbing, and when I felt better, I became careless. So don't count on me to give you wise advice!

I just told myself that crimp grip wasn't my cup of tea, though I knew it already. But it pushed me to only work on my main strengths to get even better and not give a damn for the rest. So I said goodbye to rock grinding.

The greatest problem is that the higher you get, the more invincible you feel, and the only thing you want is to reach the sky and merge in its magic.

At that point comes the question of natural abilities since at their birth, each individual isn't provided with the same physical abilities. This matter still remains a mystery, and no one can explain why some athletes, whatever the sport, are always injured whereas others are never.

I personally feel that nature has been good to me. Well, I'm not boasting about it, but I just try to make the most of it. Sure I don't drink over the top—Pastis or whatever—I've never smoked either, but it doesn't mean that my life is the one of an ascetic.

Actually, I just feel my life, and that's the first thing I realized when I started climbing. Some days, I didn't feel good; I felt like I was nailed to the ground, so each time I feel this way, I pack my stuff and do something else.

Indeed, it's great to have other interests such as a family, a pleasant job, studies, music, reading, or anything else.

Whatever the genetic legacy, I really think that sometimes, you have to listen to your body even if your mind says no. Because deep in my heart, I know that climbing is a real addiction. Without my weekly fix, I become unbearable, but this fix has to be a pure pleasure and not a fight against myself.

Obviously, today's training has completely changed, including a higher intensity in the repetition of traumatizing movements, especially on resin.

I personally don't think that today, we are climbing much more than 20 years ago, when we happened to climb nonstop for several months nearly every day—of course when the weather was fine!

I believe that you have to smile at life and enjoy the present time because no matter how careful you may be, anything can happen at any time, so seize the day!

Jacky Godoffe

Acknowledgements

Thanks to all the people who took part in that project.

Equipment

Céfar Compex Society and Pascal Adam, scientific director.
Snap Climbing Society and Fred Noé in particular.
Eb Society and Fred Tuscan, manager.

Reading-correction

Émilie Verdier, Laurent Vigouroux -extensors-, Magali Chabod, Fred Noé – electro-stimulation-, Pascal Adam -electro-stimulation-, Cécile Martha, Evelyne Gnecchi, Marie-Françoise Sembresq, Christine Parano.

Translation

Lucette Cayol

Photos

Fred Noé, Laurent Vigouroux, Dominique Thomas MCMK, Loïc Gaidioz, Jacky Godoffe, Rémi Samyn, Juliette Danion, François Lombard, Michel Azabant, Mélanie Son, Ludivine Harmand, Fred Labreveux, Marc Daviet, Violaine Beuque, John Evans.

Interviews

Jacky Godoffe, Loïc Gaidioz, François Lombard, Émilie Verdier, Mélanie Son, Juliette Danion, Rémi Samyn.

Photo 2 Targasonne boulder area

Contents

Chapter 1
Introduction

Rock climbing is a new sport which keeps developing. The increasing number of climbers in climbing facilities accounts for that phenomenon.

This enthusiasm has led to an increase in hand injuries – the hand being obviously the most exposed part of the body in climbing and consequently the most fragile.

Fingers are not supposed to be used so harshly and roughly. The intensive repetition of specific tough movements may lead to progressive or sudden traumas or microtraumas on the different structures of the hand.

According to Salomon and Vigier (1989), "climbing is a quadruped activity whose aim is to reach the top of a more or less vertical and various support or the end of the route, moving up and using only the available holds on the support, all that in a safe way."

In all categories – bouldering and difficulties – whether on sight or after work, performances keep rising. In 1980, Patrick Edlinger was successful – 7B+ – on his first try at "La polka des Ringards" in Buoux.

Nowadays, 8c is currently achieved by men. Climbers don't hesitate to go around the world in order to find or try new routes especially the most extreme ones.

These performances put fingers under a constant and intensive strain which sometimes leads to inadequate climbing habits such as the lack of stretching or warming up.

The quest for performance implies taking risks. But those risks may be reduced if the sportsman learns how to be careful and changes some of their habits. Indeed, the only way to lower the occurrence potential injuries is to be well informed and get into good habits.

The opening of several climbing facilities has led to the emergence of a new public which just like the one usually attending sports gyms comes over to practice on their free time. And due to their limited time, the new members regard warming up as a waste of time.

S. Gnecchi, F. Moutet, *Hand and Finger Injuries in Rock Climbers*,
Sports and Traumatology, DOI 10.1007/978-3-319-16790-9_1

Photo 1.1

This new behavior is the cause of potentially more serious lesions. Indeed, refusing to respect the basic principles of any physical activity – a progressive warming up, stretching exercises, good hydration, and gradual practicing after a long-term stop – may generate digit disorders.

This book is not only a source of information for climbers but also a means to make people aware of the peculiarity of rock climbing.

It also aims, at first, at describing the specificities of the lesions, then learning how to adjust one's way of life to improve healing, and finally making the return to practice easier. Understanding the doctor's diagnosis is another interesting point to focus on.

The training has to be personalized, and the recommendations have to be adapted to each sportsman or sportswoman.

The chosen examples of sportsmen going back to training are part of an empirical reflection though resulting from a mutual research between scientists and sportsmen. This is why the content of the book can't be taken for granted even if most of the fundamental principles are respected.

We accept no responsibility for recurrences or new lesions once back to practice.

In fact, the most credible information that will be passed on to climbers is a better knowledge of their own body and the search for pleasure.

The purpose of this book is absolutely not to urge climbers to make their own diagnosis and treatment. Actually after a lesion, a sportsman has to see a doctor immediately. A hand specialist would be more appropriate, since they will be the most competent to make a precise diagnosis of the lesion and prescribe the required cure.

Medical and paramedical staff is the other topic developed in this work. Far from the idea of giving them advice, they need to be informed about the particularities of this activity.

Indeed, in rock climbing, according to the posture, the grip, or the handhold, the strain felt by the finger bones may change. Consequently, it's difficult to understand and give technical or physical information to sportsmen who go back to training after a hand injury if the staff is not aware of the specificities of rock climbing and more particularly the strain constantly endured by the hands.

1.1 Different Terms Will Be Used Throughout This Publication

- The posture is the position of the fingers. In rock climbing, two main postures are possible: the open hand posture or the crimp grip.
- The grip is the way the climber clings to the hold – with two-finger or one-finger pocket for instance.
- The hold is the part used by the climber to keep going upward. Holds can be small edges or buckets. They can also be added supports in climbing gyms, CG.

We hope that climbers and coaches will find useful information on rock climbing in this book. Knowing how the body and the hand bones work, climbers will become aware of the fragility of their "handy tool" and will eventually change their climbing habits reducing likewise the risks resulting from their practice.

Photo 1.2

Photo 1.3

Chapter 2
The Anatomy of the Hand

2.1 Bone Structure

Photo 2.1

S. Gnecchi, F. Moutet, *Hand and Finger Injuries in Rock Climbers*,
Sports and Traumatology, DOI 10.1007/978-3-319-16790-9_2

2.1.1 Structures

The hand is connected to the forearm through the carpal area forming the wrist.

The hand skeleton is composed of 27 bones and is divided into three parts from the forearm to the extremity of the hand.

- The carpals which form the skeleton of the wrist
- The metacarpals which form the palm
- The phalanges which form the fingers

The hand has two sides: the palm and the dorsum.

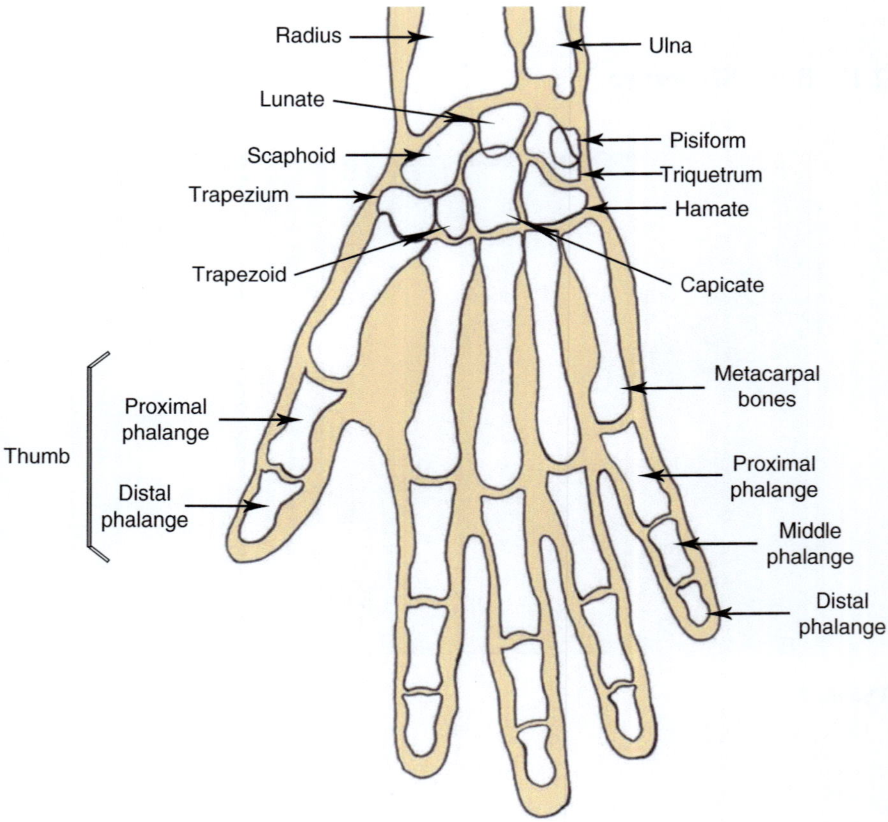

Schema 2.1 Hand bones: palmar side, right hand

2.1.1.1 The Carpals

The carpals include eight bones divided into two areas:

- The proximal area which is composed of the scaphoid, the semilunar bone, the triquetrum, and the pisiform bones
- The distal area which is composed of the trapeze, the trapezoid, the capitate bone, and the hamate bone

Due to their constitution – a transversal bending with an ulnar-radial direction, concave on the palm, and convex on the back of the hand, covered by flexors and extensors – the carpals are quite mobile.

2.1.1.2 The Metacarpals

The palm is structured by five long bones which are numbered from one to five starting from the thumb metacarpals or the first finger, which is also the most mobile of these five bones.

Each metacarpus is divided into three parts:

- The base in the proximal area
- The body
- The head in the distal area

The metacarpal base is slightly cuboid in shape. The proximal face is connected to one or several bones from the carpal distal area.

The metacarpal head is rather circular and is made of spongy bones covered with cartilage.

The metacarpal body is like a triangular prism made of compact bone tissue.

2.1.1.3 The Phalanges

Each finger is divided into three phalanges except for the thumb with only two phalanges. The hand is consequently composed of 14 phalanges including:

- Five proximal phalanges (P1)
- Four middle phalanges (P2) for the long fingers
- Five distal phalanges (P3/P2) for the thumb

Their size gradually decreases as you get closer to the distal area, and just like the metacarpals, their structure includes a base, a body, and a head.

The first phalange (P1) is called the proximal phalange, the second one (P2) middle phalange, and the third one (P3) distal phalange.

These long bones also have a convex back side and a concave palm side. The third phalange ends up in the phalangeal tuft.

2.1.1.4 The Thumb

The thumb has to be regarded as a proper finger whose main function is to create an opposition in order to act just like digital pliers.

The first phalange can move according to two axes: getting forward or backward the palm or the fingers. Contrary to the first phalange, the second phalange, just like the other fingers, is limited to one movement.

2.1.2 Finger Mechanics

When something is seized, the fingers are the first to move – clinging to the object – before being finally followed by the thumb.

The movement is reversed when something is released – the thumb relaxes first.

In a rest position, the fingers are folded in an inward arch. The little finger which is more convex than the index is also directed toward the thumb.

When fists are clenched, the first finger to fold back is the little finger followed by the ring finger and so on. The movement is reversed when the hand is opened. In fact, each unfolded finger tends to set up a chain reaction.

2.2 Joint Structures

A joint is a sliding area between two bones. Smooth surfaces such as the joint cartilage and the oily liquid – synovia – make the sliding easier. As regards the two bones, they are connected to each other by ligaments.

The finger is joined by three mobile segments. Its position results from a complex balance between several systems:

- The passive control – joints and capsular ligaments
- The semi-passive and multiple active control – superficial common flexors (FDS) and deep common flexors (FDP)

Therefore, any lesion in that complex system may lead to a dysfunction in the finger.

As regards finger folding, three joints are involved:

- The metacarpophalangeal joint (MCP)
- The proximal interphalangeal joint (PIP)
- The distal interphalangeal joint (DIP)

Their actions differ according to the need. In that precise case, MCP is used up to 77 %, PIP 20 %, and DIP 3 %.

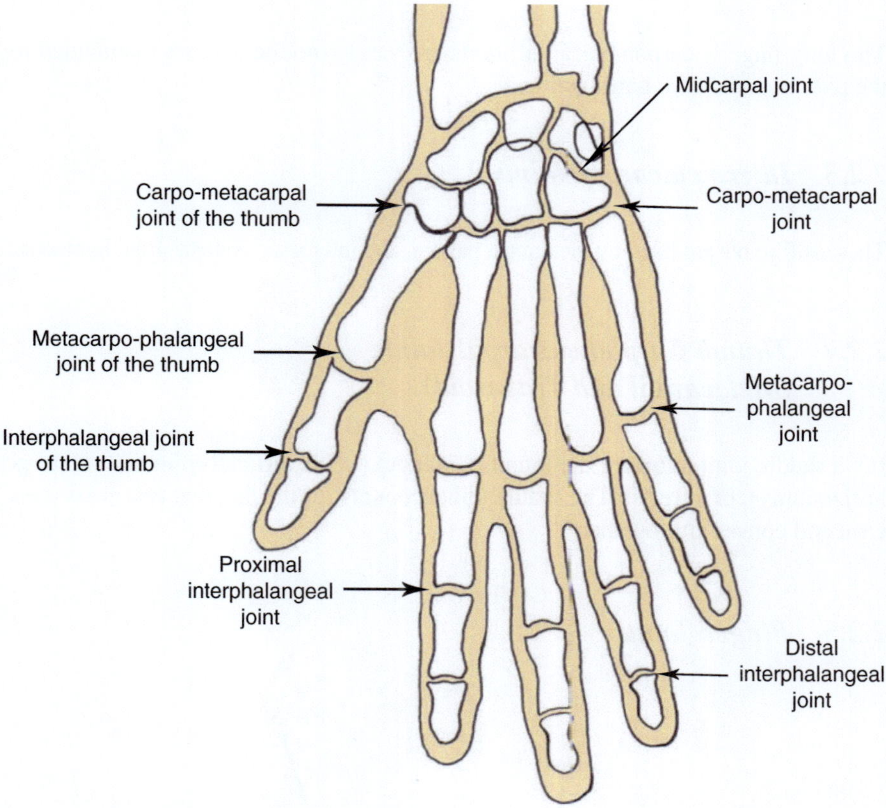

Schema 2.2 Joints: palmar side, right hand

2.2.1 Mediocarpal Joints

These joints connect the three main bones of the proximal area (scaphoid, Lunate, and triquetrum bones) to the four bones of the distal area (trapeze, trapezoid, capitate, and hamate).

A thin and loose capsule is mainly present in the dorsal side, whereas it is more powerful in the front.

Palmar, dorsal, and lateral ligaments keep the whole structure stable.

2.2.2 Carpometacarpal Joints

The long finger's carpometacarpal joints are barely mobile and are maintained by the palmar and dorsal hard ligaments.

2.2.3 Intermetacarpal Joints

These stiff joints are fixed by the dorsal, palmar, and interosseous metacarpal ligaments.

2.2.4 Thumb Carpometacarpal Joints (Metacarpal and Trapezoid)

It's a saddle joint allowing the thumb to abduct (to go vertically), to adduct (to go horizontally), or to resist. The saddle joint accounts for the fact that two areas (concave and convex) fit together.

2.2.5 Finger Joints

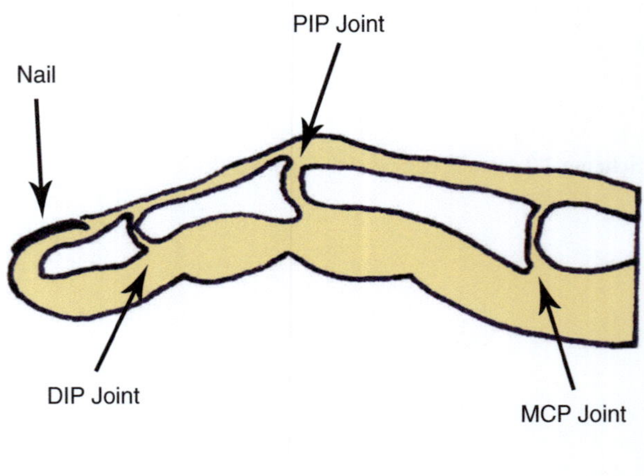

Schema 2.3 Finger joints

2.2.5.1 Metacarpal Phalangeal Joints (MCP)

The metacarpals' heads and the first phalanges' bases are the articulated elements.

The lateral ligaments reduce mobility. These joints are surrounded by loose capsules reinforced by palmar ligaments and fiber cartilages.

The MCP joint (condyloid type) is able to move laterally or vertically. The four fingers can follow two axes: flexion extension or abduction (the fingers are wide apart) and adduction (the fingers are close together). Abduction movements are practically impossible with flexed fingers.

2.2.5.2 Proximal Interphalangeal Joints (PIP)

This joint belongs to the hinge joint type – pulley joints – and can only execute a flexion/extension movement on a sagittal plane.

The PIP roughly looks like the knee joint. Lateral ligaments, radial collateral ligament (RCL), and ulnar collateral ligament (UCL) stabilize the joint. The joint capsule is thin at the PIP level and hardly interferes in the stability.

The PIP can only perform one flexion/extension movement whose average angle is 0–100°. It is also more fragile and sometimes quite tricky leading to traumas more serious than in the MCP.

Both the second and third phalanx can fold back onto the next one on a maximum angle of 90°, whereas the first phalange's degree of mobility is higher (+45° up to +90°) especially when the other phalanges are folded.

2.2.5.3 Distal Interphalangeal Joints (DIP)

This joint connects the second phalanx (P2) to the third one (P3) and presents the same constitution of the PIP.

2.3 Muscle and Tendon Structures

2.3.1 What Is a Tendon?

It's usually a white circular fiber ending. At each contraction, this resilient non-stretchable fiber, used to fix the muscle on the bone, allows the latter to move. Tendons are extremely long since the muscles at the origin of each movement are situated at half distance of the proximal area of the forearm.

2.3.2 *Various Muscles of the Hand*

2.3.2.1 Finger Flexors

Flexor Digitorum Superficialis (FDS)

This muscle comes from two points:

The humeral fascicle of muscle fibers (the main source) which ends up in the epitrochlea, that is to say, the lower part of the humerus, and in the ulnar coronoid process of the ulna

The radial fascicle of muscle fibers which follows the inner side of the radius

After merging, these two points form a unique and identical muscle at 1/3 of the forearm. This muscle is composed of vertical fibers which extend into four tendons before ending on the anterior side of the second phalanx (P2) of the four fingers. The purpose of this muscle is to flex the second phalanx on the first one, which will automatically lead to a movement in the hand and thereafter the forearm.

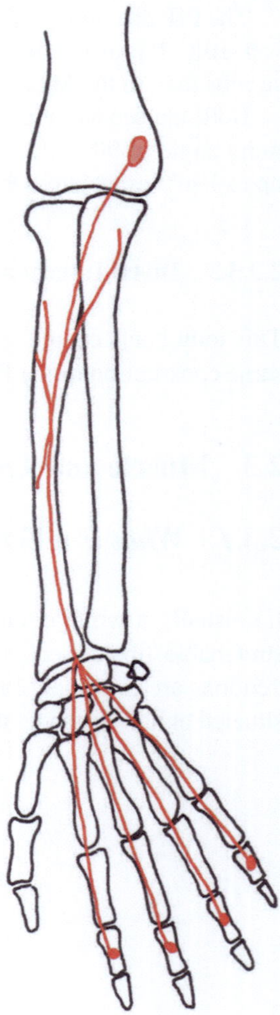

Schema 2.4 Flexor digitorum superficialis (palmar view) (Illustration inspired by Kahle (W.), Leonhardt (H.), Platzer (W.), Anatomie 1, Appareil locomoteur, 1996, Flammarion)

Flexor Digitorum Profundus (FDP)

This muscle starts below the epitrochlea on the anterior side of the ulna. These fibers are vertically oriented until they end down and turn into four tendons over the third phalanx (P3) of the last four fingers.

The purpose of this muscle is to flex the third phalanx (P3) on the second one (P2) which will automatically set up a chain reaction (from P2 to P1, from the fingers toward the hand and finally to the forearm).

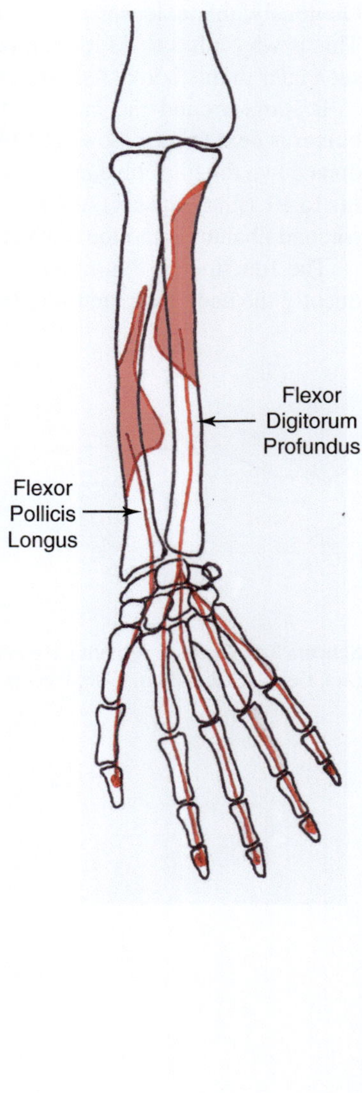

Schema 2.5 *Hand muscles* (*palmar aspect*) (Illustration inspired by Kahle (W.), Leonhardt (H.), Platzer (W.), Anatomie 1, Appareil locomoteur, 1996, Flammarion)

Flexor Pollicis Longus (FPL)

This muscle takes its source on the anterior side of the radius. It is vertically directed and ends up down the second phalanx (P2) of the thumb into a tendon.

Its function is to flex the second phalanx which will then bend the first phalanx and finally the whole thumb.

2.3.2.2 Finger Extensors

Extensor Digitorum Communis (EDC)

Obviously, this extensor muscle is certainly the most important one in climbing. This is why it has to be reinforced especially during training practice, as it will be seen later in this book (Assessment and Safety).

It starts around the lateral epicondyle of the humerus – exterior side of the humerus next to the elbow. The fibers are vertically oriented and turn into four tendons down the third phalanx (P3). It also consists of four narrow strips: the first one fits to P1 (first phalanx), the second and the third one fit to the lateral sides of P2 (second phalanx), and the fourth one fits to the base of P3 (third phalanx).

The function of this muscle is to extend P1 over the metacarpals and consequently the hand over the forearm.

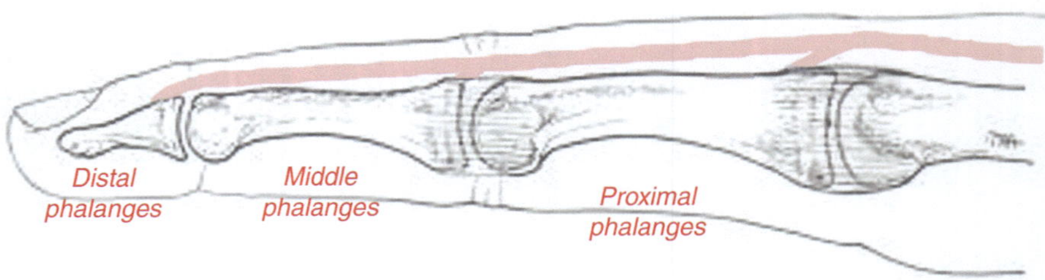

Schema 2.6 Extensor digitorum communis insertions (Illustration inspired by Bonola (A.), Caroli (A.), Celli (L.). La Main, 1998, Piccin)

Schema 2.7 Hand muscles (Dorsal Aspect) (Illustration inspired by Kahle (W.), Leonhardt (H.), Platzer (W.), Anatomie 1, Appareil locomoteur, 1996, Flammarion)

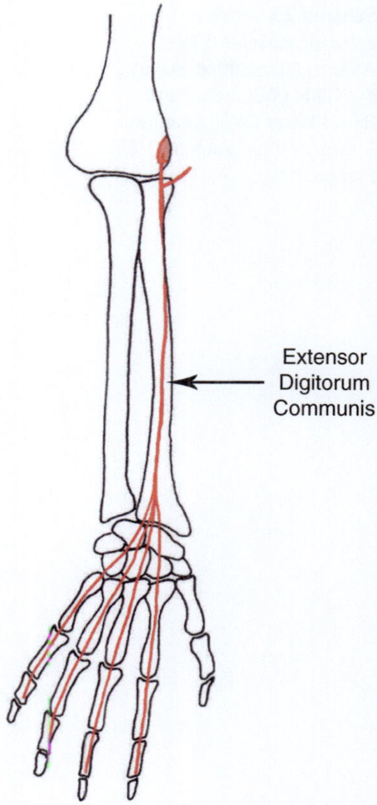

Extensor Indicis Proprius (EIP)

This muscle begins on the lower third part of the posterior ulna. The fibers are obliquely directed and end up into a tendon which lays over the finger common flexor dedicated to the index.

This muscle aims at extending the second finger.

Schema 2.8 *Fingers extensor muscles* (Dorsal Aspect) (Illustration inspired by Kahle (W.), Leonhardt (H.), Platzer (W.), Anatomie 1, Appareil locomoteur, 1996, Flammarion)

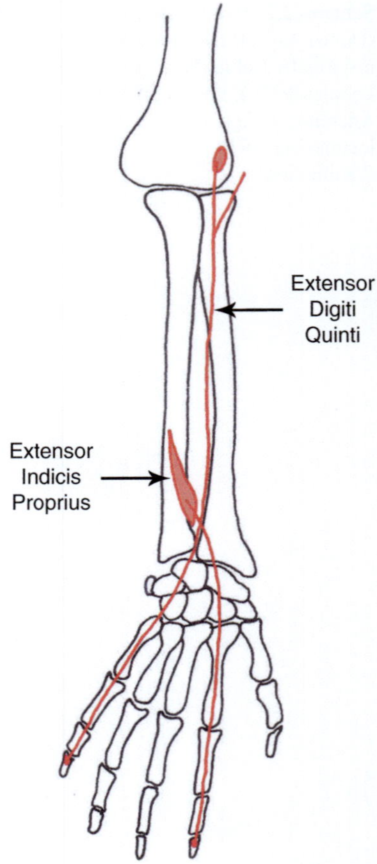

Extensor Digiti Quinti (EDQ)

This muscle originates around the lateral epicondyle of the humerus. The muscle fibers are diagonally oriented and eventually form a tendon which merges with the fifth finger common extensor.

Its purpose is to extend the little finger which might be quite necessary in climbing, especially when the climber wants to improve the hold by using a grip without the little finger.

Extensor Pollicis Brevis (EPB)

This muscle comes from the posterior side of the radius and the ulna before ending in the first phalanx (P1) of the thumb. The fibers are obliquely directed.

It aims at extending the thumb P1.

Extensor Pollicis Longus (EPL)

This muscle starts around the posterior area of the ulna and ends up in the second phalanx (P2) of the thumb. The fibers are obliquely directed.

It aims at extending the thumb P2.

2.3.2.3 Abductor Muscles

Abductor Pollicis Longus (APL)

This muscle originates from the posterior side of the radius and the ulna before ending in the first metacarpal with obliquely directed fibers.

Its purpose is to abduct with the thumb. In climbing, its grip function is a priority since it allows the thumb to be used as "pliers."

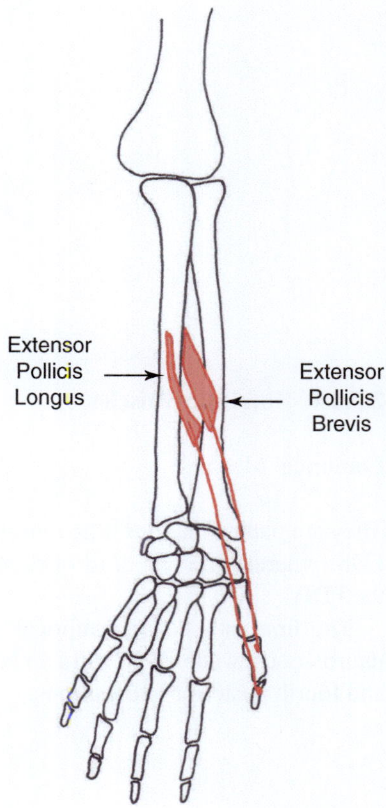

Schema 2.9 Muscles of the hand (dorsal aspect) (Illustration inspired by Kahle (W.), Leonhardt (H.), Platzer (W.), Anatomie 1, Appareil locomoteur, 1996, Flammarion)

Schema 2.10 Abductor pollicis longus (dorsal aspect) (Illustration inspired by Kahle (W.), Leonhardt (H.), Platzer (W.), Anatomie 1, Appareil locomoteur, 1996, Flammarion)

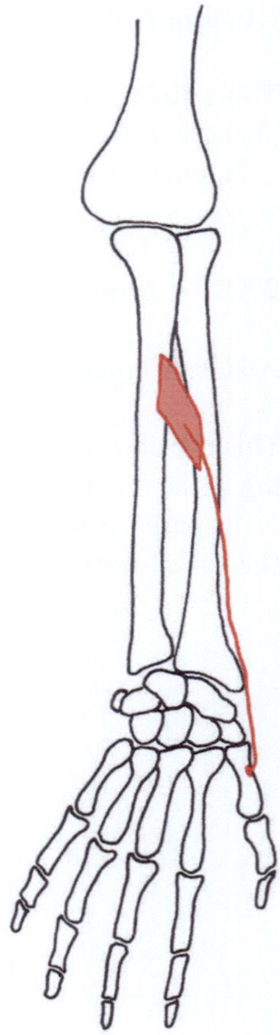

2.3.2.4 Intrinsic Muscles

Lumbrical Muscles

They are part of the four little muscles. The first one begins at the lateral side of the FDS, whereas the last third ones merge with the two tendons not far away from the FDP.

The first and second lumbrical muscles end up in the first and second dorsal interosseous, whereas the third and the fourth lumbrical muscles end up in the third and fourth posterior interosseous.

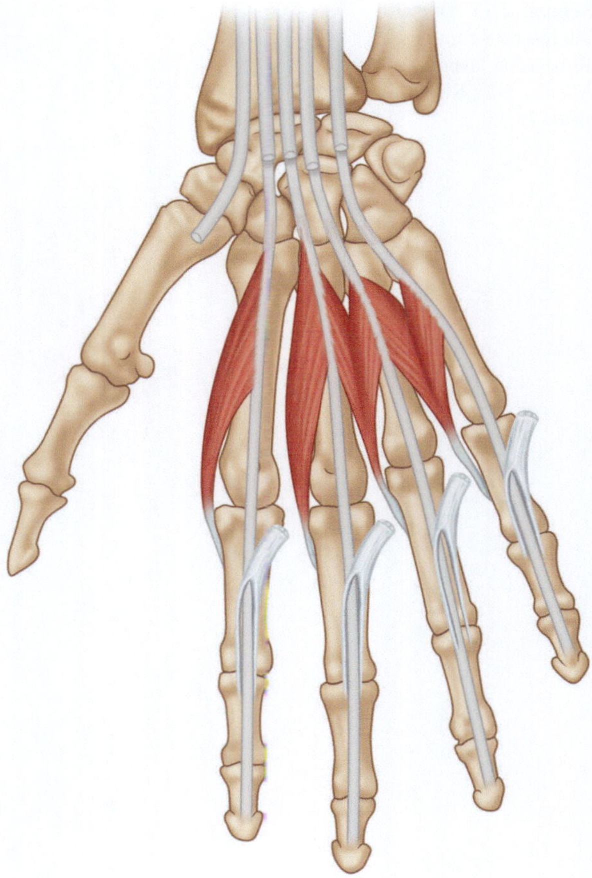

Schema 2.11 Lumbrical muscles (Illustration inspired by Bonola (A.), Caroli (A.), Celli (L.). La Main, 1998, Piccin)

The purpose of the lumbrical muscles is not only to flex P1 but also to extend P2 and P3. They are a means to regulate the tensions between the flexor system and the extensor system, but they are also a good help to adjust the grip in a more precise way.

Dorsal Interosseous Muscles

These muscles are situated between the metacarpals and converge on the third finger. They fit into the lateral sides of the adjacent metacarpals.

They are part of the adduction movement – getting the fingers close to each other – and the extension of the third phalanx (P3).

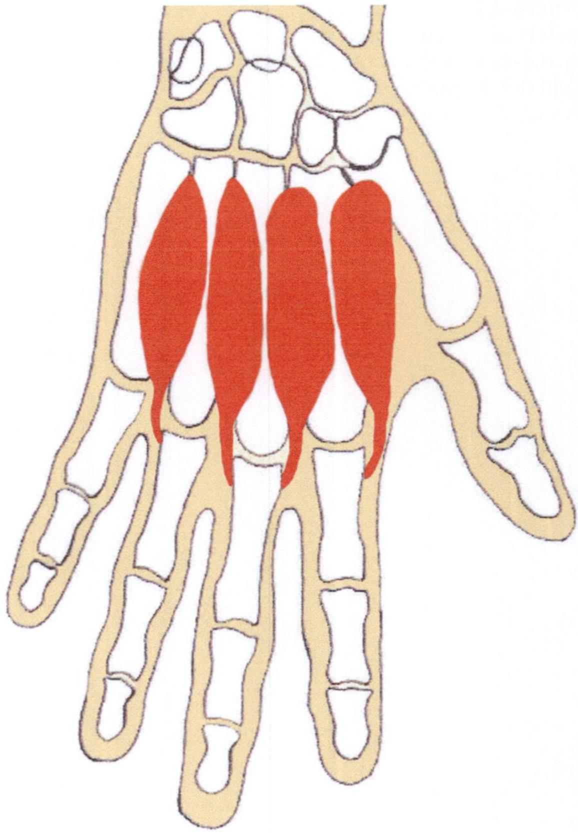

Schema 2.12 Dorsal interosseous muscles. Illustration inspired by Michael (L.), Richardson (M.D.)

Volar Interosseous Muscles

These muscles are located between the metacarpals and diverge along the third finger. They are attached to the lateral sides of the appropriate metacarpals.

They are part of the abduction movement – getting the fingers apart from each other – and the flexion of the first phalanx (P1).

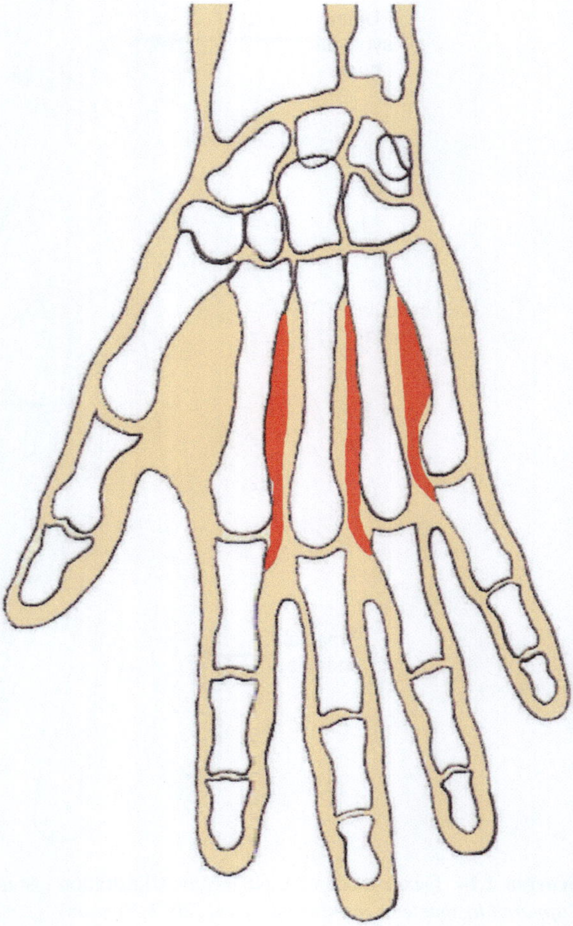

Schema 2.13 Volar interosseous muscles (Illustration inspired by Michael (L.), Richardson (M.D.))

2.4 Synovial Sheaths

Some tendons are enveloped by what is called a synovial sheath. Several synovial sheaths can be found in the hand:

- The radial synovial bursa envelops the tendon of the thumb's long flexor muscle.
- The ulnar synovial bursa, which is much larger, envelops the tendons of the four long fingers' deep and superficial flexors, before shrinking around the little finger flexor.
- The index finger, the middle finger, and the little finger have their own sheath.

All these sheaths play an important part in the good sliding and nutrition of the sinews thanks to the synovial liquid.

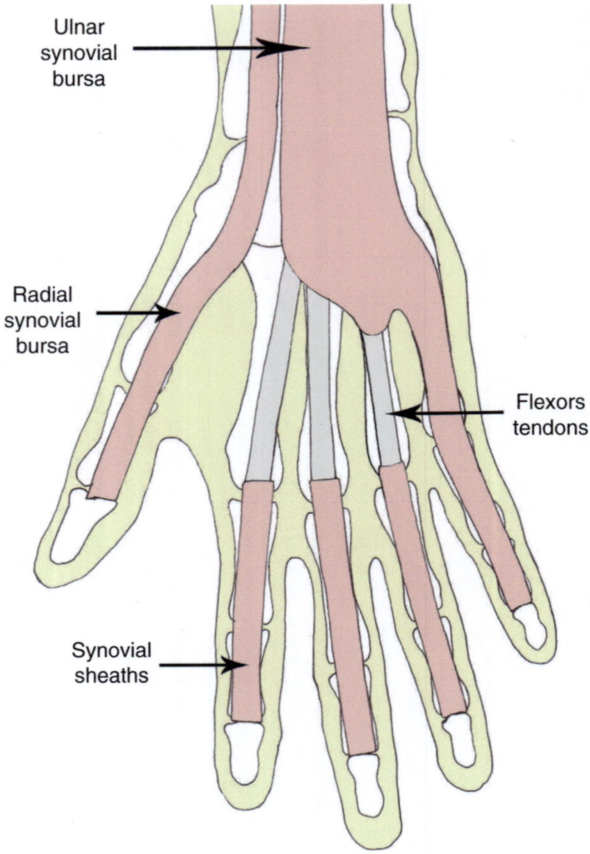

Schema 2.14 Flexor tendon digital sheaths (Illustration inspired by Dufour (M.), *Anatomie de l'appareil locomoteur*: *membre supérieur*, 2002, Masson)

2.5 Ligament Structures

A ligament is a silver-white network of fibers which are extremely tight, hardly expandable, and difficult to cut off. They are supposed to keep the right balance of the articulations and restrain the movement to a certain point.

In the following part, due to the complexity of the carpal joints, the focus will be limited on the joints of the fingers.

2.5.1 MCP Joints

Several ligaments reinforce the metacarpal phalangeal joint: collateral ligaments, the intermetacarpal transversal ligament, and the palmar ligament (volar plate).

The collateral ligaments are tensed during a flexion and loose during an extension. Therefore, moving laterally (abduction or adduction with two fingers) is impossible during a flexion.

The intermetacarpal transversal ligament is a white narrow fibrous band covering the head of the second, third, fourth, and fifth metacarpals.

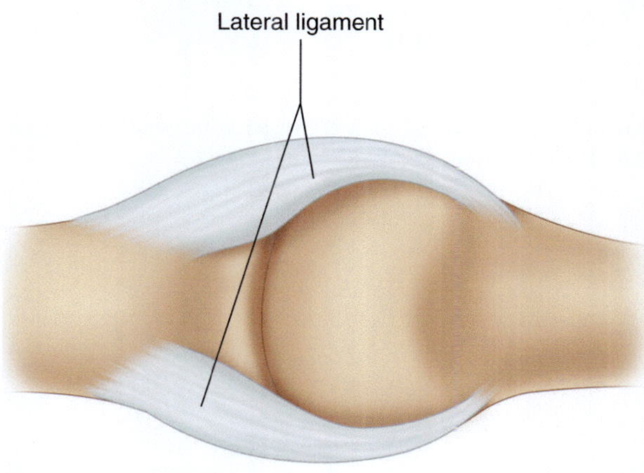

Schema 2.15 Metacarpophalangeal joint (palmar aspect) (Illustration inspired by Kahle (W.), Leonhardt (H.), Platzer (W.), *Anatomie 1, Appareil locomoteur*, 1996, Flammarion)

2.5.2 PIP Joints

Collateral ligaments include two networks:

- The main network (RCL) which is on the lateral side of the head of P1. It spreads out all over the lateral side of the base of P2, which accounts for the fact that it is constantly tensed (in flexion or in extension) and consequently enhances the lateral stability.
- The secondary network (UCL) which has the same source than the RCL but ends up on the lateral side of the volar plate. Just like the latter one, it loosens in flexion. Collateral ligaments seem to be the key of the frontal stability of the PIP.

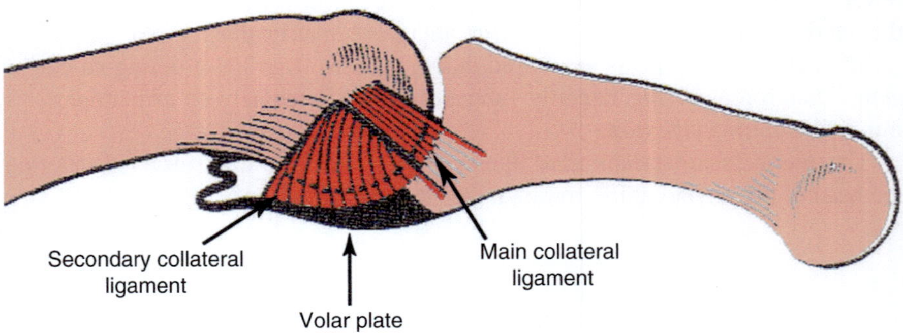

Schema 2.16 Proximal interphalangeal joint (Illustration inspired by Tubiana (R.), *Traité de chirurgie de la main*, 1980, Masson)

The palmar ligament or volar plate is a resisting and thick fiber cartilage in which its deep side is covered by cartilage extending likewise the glenoid cavities of the P2 base. The former one is present throughout the width of P2 and reveals on its side a powerful common area with the main network of the lateral ligament and two lateral expansions which reinforce A2 pulley. The two functions of the palmar ligament are first to limit hyperextension and secondly to direct properly the flexors.

2.5.3 DIP Joints

The ligaments of the DIP are quite similar to those of the PIP.

2.6 Fibrous Structures: Digital Pulleys

Pulleys are inextensible fibrous structures, creating an osteofibrous canal called "digital flexor tendon digital sheath." They are meant to hold tendons in contact to bones to improve the mechanic system of the flexors and consequently to avoid a bowing string effect during a flexion.

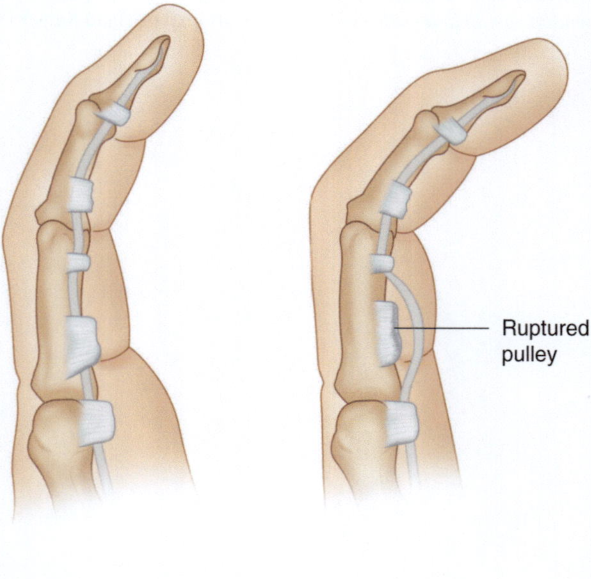

Schema 2.17 Digital pulley injury (Illustration inspired by Klauser et al., *Finger Pulley Injuries in Extreme Rock Climbers. Depiction with Dynamic US*, 2002, Radiology)

Several pulleys are present on each finger:

- The palmar aponeurosis pulley (AO or PA) which is fixed on each side of the synovial sheath.
- The annular finger pulleys (A1, A3, and A5) which are, respectively, inserted to the volar plate of the MCP, PIP, and DIP articulations. A2 and A4 are thicker and fit at the periosteum level of P1 and P2 phalanges.
- The cruciform pulleys (C1, C2, and C3) which are less thick, suppler, and between the ring finger pulleys. They fold back during the flexion of the finger. Likewise, C1, C2, and C3 cruciform pulleys make the connection between the ring finger pulleys – A2/A3 for C1, A3/A4 for C2, and finally A4/A5 for C3.

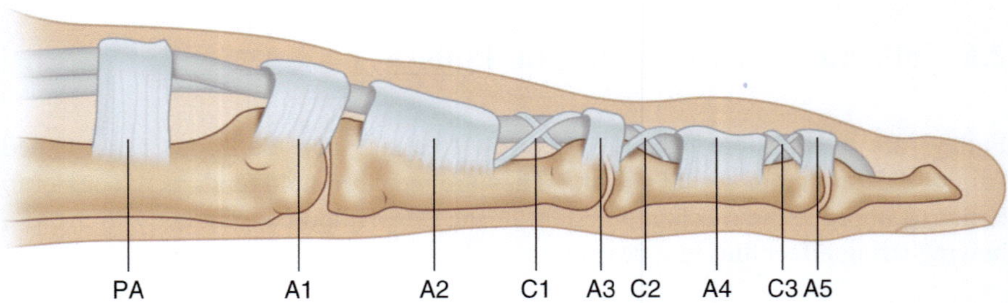

Schema 2.18 Digital pulley anatomy (Illustration inspired by Doyle Jr., *Anatomy of the flexor tendon sheath and pulley system*, 1988, Journal of Hand Surgery)

When A2 and A4 pulleys are totally severed, complete flexion is impossible due to the so-called bow stringing phenomenon. These pulleys are the most important regarding biomechanics; consequently, contrary to the other pulleys, which are more flexible and less robust, the former ones can be easily damaged. So to sum it up, A1, A3, A5, and cruciform pulleys are much more flexible and are far less tough than the former ones (A2 and A4).

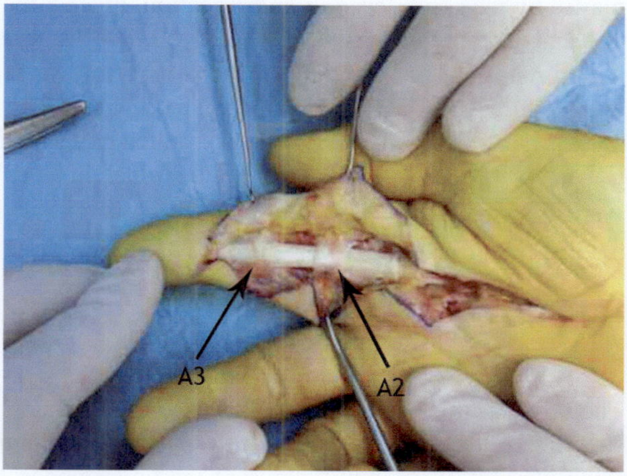

Photo 2.2 Pulleys' operating view

2.7 Structural Adaptations

2.7.1 Climber's Hand Skin

The skin is not only a protective and thermoregulation organ but also a means to socialize or acknowledge one's belonging to an ethnic group (according to the color).

As far as this book is concerned, the most important thing to know is the part played by the skin especially in perception. Indeed, the skin is an organ of perception and touching finds its origin in pulps which are spread throughout the skin. The digital pulps are filled with mechanical receptors gathering various information as temperature, pressure, tactile discrimination, etc. The skin of the dorsum of the hand is hairless, thinner, and suppler than the palmar skin of the hand. This flexibility allows the hand and the fingers to be folded in an easy way.

Climbers' hand skin tends to adjust to the specificities of climbing, which consequently leads to quite visible changes.

2.7.1.1 Skin Cover

The skin gets thicker due to a hyperkeratosis in the pulp area (roughness of the most superficial layer called cornea) with a chronic inflammation of the hyponychium (the pulp area situated at the end of the finger under the nail) and the lateral line around the nail. This phenomenon accounts for the compact and gnarled look of the climber's hands, which are constantly gripping or rubbing while climbing. Indeed, the perpetual brushing over stiff edges gradually tans the skin and leads to changes which may alter the skin cover and be the cause of peripheral hyperkeratosis.

The perpetual use of magnesia tends to make the skin dryer and consequently enhances the skin deformity all along training practices.

Despite the usual examples mentioned above, some other insignificant lesions may appear on the skin.

2.7.1.2 Sensitivity

According to the two-point discriminative test (Weber test), two high-level climbers (8a or more) out of ten have seen their sensitivity reduced. This fact is obviously due to the skin growing thick after years of practice.

2.7.1.3 Digital Volume

The digit perimeter is quite high especially when it's measured between the proximal phalanges (P1), the intermediary ones (P2), and the PIP joint. In fact, the perimeter is always high in the DIP area. This phenomenon was proved by a blind test done in Spain between two groups of climbers and supervised by Luis Aguilella d'Alzira.

Indeed, as regards the climbers' group, the perimeter of P1 is always inferior to the PIP one, whereas concerning the test group, the perimeter of P1 is equal to that of the PIP. The difference is quite obvious ($p<0.0001$) and can be explained by the fact that the PIP is more protruding because of repeated capsule-ligament ruptures and a hypertrophied healing compared to the common structure. The volar plate hypertrophy visible on the MRI also enhances the largeness and the knottiness of the climber's PIP.

In fact, the whole finger is wider. And this phenomenon has been revealed after biological tests showing how the periosteum had grown thicker in P1proximal phalange (in six cases out of ten) and in P2 intermediary phalange (three cases out of ten). Besides, sometimes real traction diverticulum has also appeared due to the high strain endured by the pulleys (A2 for P1 and A4 for P2).

2.7.1.4 Vascularization

Regarding the vascularization of the digit pulp, a change has been noticed. Indeed, the nail groove is whiter than what is usually found in common people.

2.7.1.5 Arthritis

We've always thought that we'll be the witness of osteoarthritis (joint cartilage degeneration) in climbers' strained hands and that one day we'll have to treat them. But fortunately, up to now, even after about 30 years of consultations – for the more experimented of us in that field – this problem has never come up significantly.

To conclude, I would say that practicing climbing at a high level leads to structural changes and adaptation transformations in climber's hands and fingers.

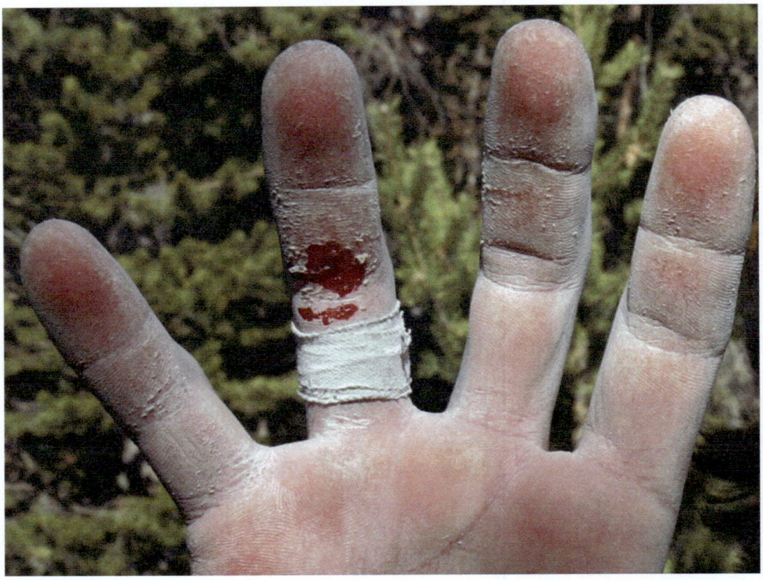

Photo 2.3 Typical aspect of overused hand: globular pulps, hyperkeratosis, and dry skin due to overuse of magnesia. The 4th finger PIP lesion is a recent cutaneous abrasion

Photo 2.4

Chapter 3
Various Injuries

Photo 3.1 Slope grip

S. Gnecchi, F. Moutet, *Hand and Finger Injuries in Rock Climbers*, Sports and Traumatology, DOI 10.1007/978-3-319-16790-9_3

3.1 Tendon Injuries

3.1.1 Tendinopathy

3.1.1.1 Tendinitis

Introduction

Tendinitis is an inflammation of a tendon and its sheath. The clinical symptoms involve a pain, a specific or diffuse swelling, and a difficulty in the movement.

The tendon looks like a rope and is made of smooth and resisting fibers. So just like a rope, the tendon is subject to several changes starting from simple fraying to eventually the final rupture. Repeated movements, intensive efforts, a bad posture or a long-term identical position, vibrations, and a physical stress are a few examples of what may lead to an alteration in the tendon. Any change in the tendon results in an inflammatory reaction, that is to say, a local reaction in the tissue after a lesion. Throughout time, inflamed tendons become irregular, getting thicker and battered. In order to get back to normal and to avoid chronic fragility of the tendon, a break is highly advised.

The three following signs reveal the weakness of a tendon:

- Firstly, the pain blurs away during the warm-up.
- Secondly, the pain is more persisting.
- Thirdly, the pain is omnipresent even during rest periods.

Photo 3.2 Three-finger grip open hand posture

Tendinitis Factors

Several causes are responsible for tendinitis. People don't react the same way according to the intensity or effort required in climbing; this is why training has to be personalized. Tendinitis appears after a too much repeated or exaggerated work. Bearing that in mind, avoiding injuries is consequently possible as long as training main principles are respected (cf. part Training and Back to Training).

To start well, a training practice has to begin with a general warm-up before focusing on precise muscles. The intensity may vary according to the climber's capacities, but the most important is to adjust the practice to the present physical condition and level of the climber. To finish, it's a priority to stretch the muscles – get back to a relaxed posture before any effort – to reduce the tension.

What really matters is to listen to one's body and learn how to deal with the body signs telling people to emphasize, maintain, or reduce the intensity of any effort. In case of a persisting pain after the warming-up, it's a priority to stop the ongoing practice and consult a doctor.

In climbing, several factors may account for tendon traumas:

- An overexaggerated repetition of the same movement
- Holds with stiff edges
- Too small holds preventing a precise grip or the use of all fingers
- Crimp grip
- A bad posture in climbing up
- Finger reception during a grip after a dynamic movement or a dyno which is quite traumatizing for the tendon in contact with the hold

But one of the main causes of tendinitis is dehydration, which damages the resilience of the tendon tissue (cf. part Assessment and Safety). The lack of water is very common; this is why the climber has to learn how to drink before, during, and after the effort.

Another cause of tendinitis is a too high-protein diet. Indeed, proteins tend to enhance the volume of urea and uric acid which are extremely toxic for the tendons. High-protein diets may also cause muscular hypertrophy which might damage the tendon that has not undergone the same changes and still remains less resilient and less strong.

A chronic dental infection is also another factor for tendinitis, since tendons are extremely sensitive to that phenomenon. Consequently, a good dental hygiene and a regular dental checkup are a good way to avoid tendinitis not resulting from a usual practice.

So it is important to bear in mind that a bad dental hygiene may lead to a slight moderate infection, which might be nonetheless quite toxic for tendons.

Localizations

The feeling of a pain on the injured part of the body is the clinical sign of tendinitis problems.

In climbing, a hand tendinitis is most of the time located on finger flexor tendons, but muscle flexor tendons may also be injured. The finger superficial flexor is the most frequently wounded part. The pain is then located in the P2 and P1 phalanges area – due to the constant strain resulting from the crimp grip posture – more particularly in the middle and the ring finger.

Clinical Data

Typical physical signs of a tendinitis can be revealed after a medical exam:

- A painful feeling on the injured part
- A possible swelling
- A red blotch on the wounded area
- A reduced mobility

Of course, all these symptoms may not appear at the same time, but one of them can be the sign of a tendon lesion.

According to the seriousness of the tendinitis, the pain is first felt before the warming-up and gradually disappears as the warming-up starts, squeezing a ball in one's hand for instance.

The clinical exam has to be bilateral and comparative to detect surely and precisely the location and the gravity of the lesion. The former one includes several steps: checkup, contraction against resistance, passive stretching, and palpation.

The checkup is just a way to get superficial details about the tendons, such as the volume, the outline of the soft lateral parts, and the muscle's general condition.

Contrary to the checkup, the contraction against resistance exam is a fundamental key of the clinical tests. Indeed, as regards tendinitis and to confirm a proper diagnosis, it is a priority to find out if pain is felt after contraction against resistance.

The second most important exam is the passive stretching, which sometimes may be painful or with a small limitation in the amplitude of the movement. Those exams can be easily performed as long as the people concerned have a small knowledge in anatomy and physiology.

As regards the palpation exam, it has to be meticulously done, since it is a means to detect different anomalies such as:

- A spontaneous pain felt by the patient
- A locally increase in temperature
- Localized or spread friction
- A painful moderated or voluminous nodule

Most of the time, the pain is gradual, but it can be quite sudden in some areas. In fact, a direct shock, just like a usual physical activity or training, sometimes more intense or totally new, may cause, aggravate, or start up the pain again. A change in the training habits, in the technique, or in the equipment used may also trigger new pains.

Treatment and Behavior

After a painful feeling in one part of the hand, obviously, the wisest solution is to stop the practice. The most efficient and immediate treatment is to stop any activity and to put some ice or use any other cryogenic method – the use of very low temperatures. The complete stop of any practice implies the obligation not to use the inflamed limb. It is highly advised not to grip any objects with the injured finger. Pain killers can be prescribed by doctors to ease the pain.

Several icing techniques are available. One of them consists of, first, applying ice on the lesion for 10 min, then waiting for 20 min to let the blood circulate and drain the injured part, and finally icing again for ten more minutes. This operation needs to be repeated three times a day for 3 days. Cryogenics is very efficient in inflammation cases, and the topic will be thoroughly dealt with in the Chap. 6.

According to the seriousness of the lesion, several additional creams can be used to ease the pain. Nonsteroidal anti-inflammatory creams (NSAI) such as Ketum®, Nifluril®, or Voltarene® are quite efficient once locally applied. But to cure the most serious cases, cryogenics, electrotherapy (treatment using the electric current), and mesotherapy (locally simultaneous injections of anti-inflammatory drugs) are necessary. According to the injections, infiltrations can soothe the pain (pain-killer effect) or fight against inflammation (anti-inflammatory effect).

Therefore, no strain has to be undergone by the injured limb which is not healed yet even if the pain has been killed, that is why infiltrations are not recommended for finger lesions.

Other means are available to help tendon healing. The deep transverse massage (DTP) is a to-and-fro motion which "irritates" the tendon and consequently improves vascularization. This massage consists of making small movements transversally to the tendon or on the edge of the tendon-bone insertion. After a while, if the strain is tolerated, the pressure will be enhanced, and instead of using the finger pulp, the edge of the flexed index medium joint will be used.

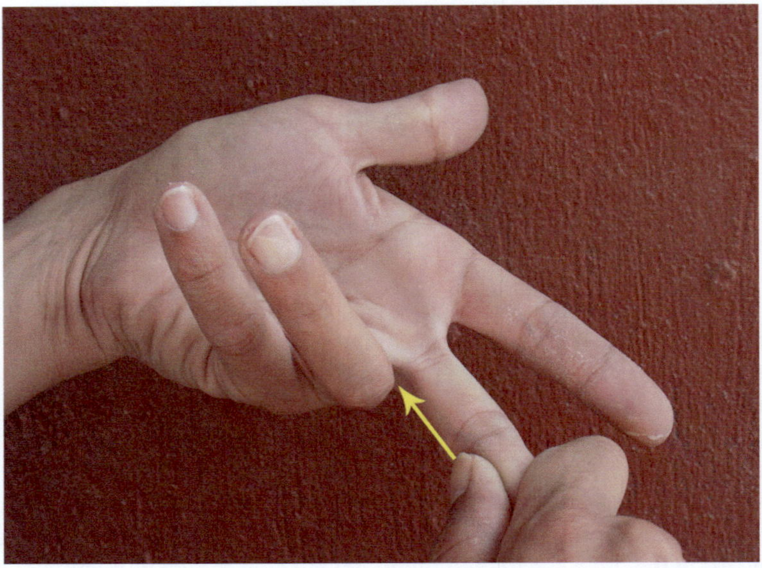

Photo 3.3 Transverse massage

A tendon is part and parcel of a chain reaction, which can be slightly disturbed by a dysfunction in the vertebra, back muscle, or anything else. Any small change in the tendon axis can lead to torsion and consequently to rubbing of the tendon in its sheath. Therefore, in case of serious fall in lead category, when a climber has been "spot" by a bilayer, or in bouldering, a fall on the back, the coccyx, or outside the crash pads, it is highly recommended to consult a specialist to check the integral rachidian system of the different muscular chains, in order to detect any even harmless lesions connected to climbing risks. Therefore, a potential fall at high-level climbing has to be considered, because it can have harmful side effects on other parts of the body.

After a tendinitis, full sports practice has to be interrupted for 2 weeks; nonetheless, it is highly recommended not to go back to training as long as the pain persists.

Medical Scanning

X-rays are not really helpful as regards tendon injuries contrary to MRI and scanning which are more relevant to give a proper diagnosis.

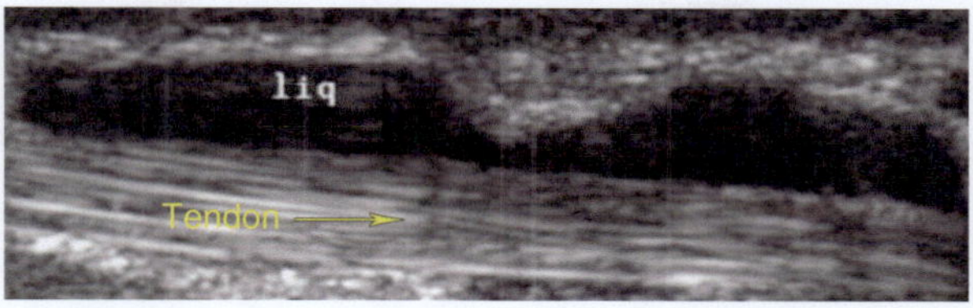

Schema 3.1 Tenosynovitis ultrasonography

3.1.1.2 Tenosynovitis

Introduction

Tenosynovitis is an inflammation of the synovial sheath surrounding the tendons. The function of this sheath is to ease the sliding of the tendon especially during a change of direction. An oily liquid – synovia – is secreted by the walls of the synovial sheath. Some factors such as microtraumas due to constant rubbing or repeated shocks can lead to a dysfunction in the lubrication system. An inflammation and a swelling of the synovial sheath can result from the constant rubbing between the tendon and the synovial sheath. When the normal sliding is hampered, a pain and a "crepitus" – friction – are felt while in action.

This injury is often quite sudden and has to be treated immediately in order to avoid the weakness of the tendon, its adhesion to the synovial sheath, or any other more harmful problems.

Tenosynovitis Factors

Joint strain is often one of the causes of tenosynovitis especially when a move is currently repeated leading to microtraumas. This tendon lesion happens when tendons are overstrained during an effort highly above one's capacities in order to go over one's limits.

Safety precautions about tendinitis have also to be taken to avoid tenosynovitis.

Localization

A tenosynovitis usually affects the finger flexor muscles, but it may also affect the extensors, though that lesion is much less common. The pain results from a long and intense rubbing especially during the crimp grip and is generally felt in the area of the P1 phalange. The pain may eventually spread over the palm up to the forearm.

Clinical Data

When a pain is suddenly felt in the area of the P1 phalange, a precise diagnosis has to be made to find the appropriate treatment and not to take this injury as a rupture of the pulley.

Treatment and Behavior

The first thing to do is to get treated, and the second step to take is stopping climbing practice. The finger needs some rest and has to be joined to the next finger. It is advised to ice the injured finger several times a day (cf. Chap. 6).

It is also highly suggested to see a hand specialist or a sports doctor who will prescribe the right treatment. As regards tenosynovitis, one has to bear in mind that infiltrations are to be avoided.

After a tenosynovitis, full sport practice has to be interrupted for 2 weeks; nonetheless, it is highly recommended not to go back to training as long as the pain persists.

Medical Scanning

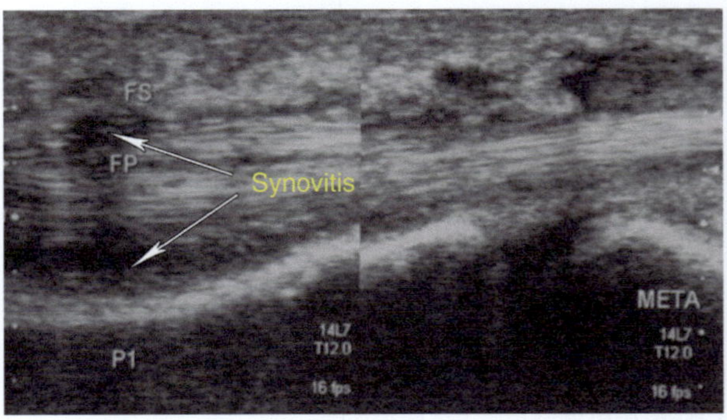

Schema 3.2 Tenosynovitis ultrasonography. *FDS* flexor digitorum superficialis, *FDP* flexor digitorum profundus, *P1* proximal phalanx

Tendinopathy Recap

Symptoms

- A painful feeling in the lesion area
- A possible swelling or red blotch
- A sudden sore

Treatment

- Icing three times a day
- Anti-inflammatory gels prescribed by doctors

Rest Period

- About 2 weeks up to 6 weeks

Behavior

- Immediate stop in climbing practice
- No more practice as long as the pain is still felt
- See a doctor

3.1.2 *Synovial Cyst*

3.1.2.1 Introduction

A cyst is an abnormal swelling filled with liquid matter and most of the time connecting to a precise joint. In climbing, as far as fingers are concerned, the tendon sheath synovial liquid can prevent finger flexion and consequently hamper hold gripping.

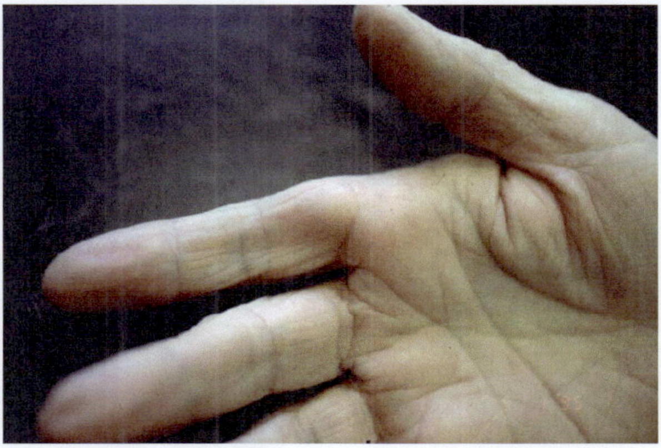

Photo 3.4 Synovial cyst (into the digital sheath) at proximal index level

3.1.2.2 Synovial Cyst Factors

The excessive use of a joint is one of the main factors of synovial cysts. The sheath inflammation may cause a nodule to occur.

3.1.2.3 Clinical Data

It is a swelling on the flexor tendon located at the basis of the finger. This swelling may be painful when touched or when involved in a movement.

As time goes by, the swelling may grow bigger and is perfectly recognized with its well-limited outline and the elasticity and the hardness of the cyst.

3.1.2.4 Treatment and Behavior

A treatment is not always necessary for the sportsman because this cyst may change and eventually disappear spontaneously. A period of rest may help, and surgery may be involved in case of long-term troubles. Other side treatments such as punctures are hardly used due to the inevitability of a recurrence. In surgery, the swelling and its basis are entirely retrieved. And usually, the results imply no problems of stiffness, even if a recurrence is always possible.

3.1.2.5 Medical Scanning

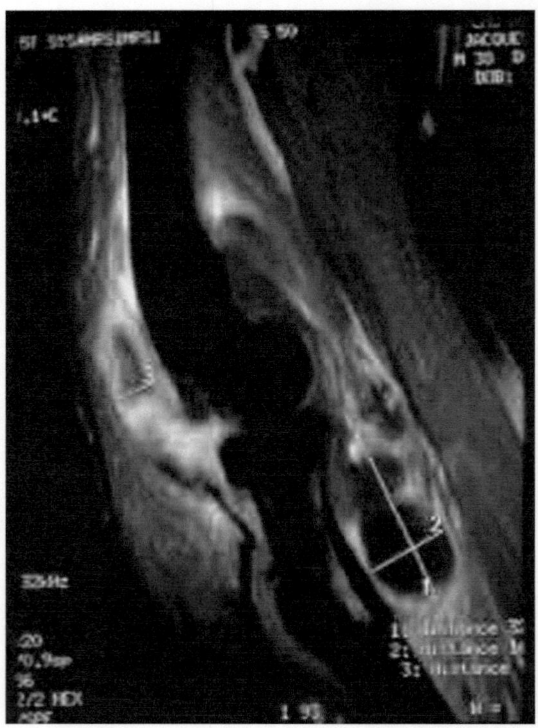

Schema 3.3 MRI: synovial cyst

3.2 Pulley Injuries

3.2.1 Introduction

In climbing, the intensive training has led to the appearance of new lesions such as pulley rupture – with or without bow stringing.

This trauma is one of the most feared by climbers since it implies a complete stop in climbing practice. Besides, very few preceding signs can be detected. And most of the time, this microtrauma affects high-level climbers. Surgery is not systematic; nonetheless, when it has to be done, 80 % of climbers go back to normal – sometimes they even achieve higher capacities.

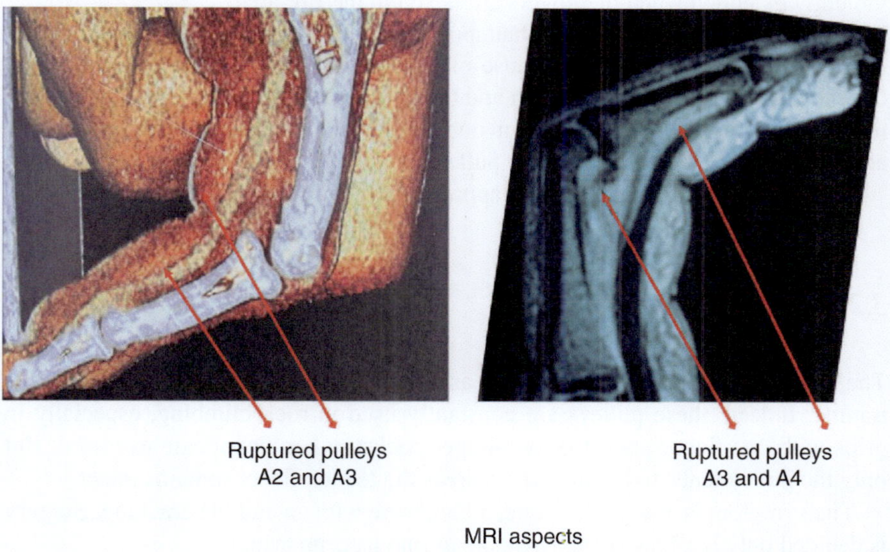

Schema 3.4 Bow stringing effect (Illustration inspired by Chevrot (A.), Clinical Imaging wrist and hand,1998 Masson)

3.2.2 Pulley Rupture Factors

Most of the time, this kind of lesion occurs during intensive training in CG, especially during a repeated work on a difficult route. Having great expectations, the majority of top climbers keep practicing, repeating on and on the same movements, and consequently straining or putting pressure on a finger or a pulley. Besides this, pressure is enhanced by a limited time to get rest between each attempt, reinforcing the strain and the weight undergone by the same finger structure. So the numerous tries before going through a difficult route successfully are responsible for localized pulley lesions. The breaking of the pulley usually happens in crimp grip during a dyno or a repeated difficult route.

In crimp grip posture, the finger joint is easily torn off. The climber's weight and the number of fingers used are factors which make the tearing complete. Whatever the cases, all the pulley breaking follows the same pattern; the PIP is suddenly and violently extended when the fingers are flexed in the PIP area.

Particular training sessions on hangboard or campus board are also quite straining for the pulleys and are the cause of lesions. During these trainings, the grip depends only on the finger strength and the climber can't use his legs to get "some relief." The body movements from one grip to another, especially in dynamic actions, cause violent shocks on the pulleys.

The lack of warming-up and an inappropriate hydration indirectly lead to lesions.

3.2.3 Clinical Data

The third- and fourth-finger pulley A2 and A3 seem to be the most affected by this trauma. Indeed, these pulleys are constantly used in rock climbing, especially in grips with one-finger pocket or two-finger pocket, when flexors are overused. But only the pulley tends to be injured, whereas the tendon flexor remains intact.

The symptom is always the same, a vivid sore with an audible cracking. Surgery is decided only if a bow string is visible during a crimp grip.

The other signs of this lesion are the following:

- A pain on the pulley during palpation
- A soreness when flexing/resisting P2 over P1
- A located edema
- An incapacity in mobility and the presence of a bow string according to the seriousness of the injury

3.2.4 *Treatment and Behavior*

The climber has to stop immediately any practice to prevent any worsening of the injured pulley or the other ones. Then icing has to be done three times a day until the hand specialist's visit. Meanwhile, the sportsman can also bandage his finger joining it to the next one – finger knitting. In any case, no strain has to be endured by the injured finger.

If there's no bow stringing, the preventive treatment is essentially based on a thermally molded protective ring. Physiotherapy and ultrasounds may complete the treatment. The rigid protective ring has to be maintained for about 45 days, and of course a complete cessation of activity is the key to accelerate the healing process. No weight loading has to be undergone. This latter behavior is probably the most difficult to accept by top competitors.

The hand specialist has to detect the presence of a bow string – bilaterally and in crimp grip. The test is positive when the doctor feels the flexor sticking out under the skin as he is auscultating the patient. When in doubt, the diagnosis will be more precise with a scanning or an MRI.

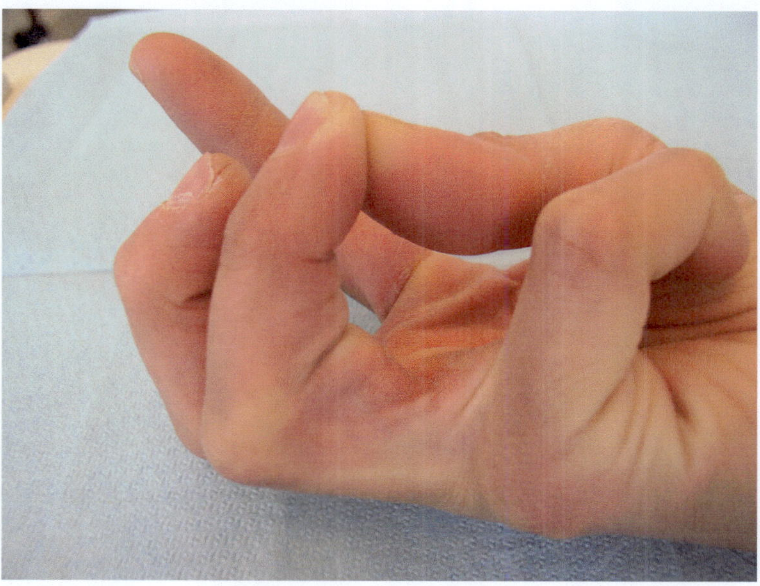

Photo 3.5 Looking for bow stringing. One can see the FDS bulging under the skin on the fourth finger at the P1 level meaning "bow stringing"

The surgery consists of recreating a structure holding back the flexor to the phalanx when the latter is in flexion. This replacing structure is an extensor dorsal retinaculum graft – CDLL graft. The reconstruction is safely protected in a wrist flexion splint with flexed MP joints for 45 days. Then, it is replaced by a rigid ring for another 45 days at the repaired site. Going back to rock climbing will be possible after 3 months, paying attention to the usual safety recommendations such as good hydration, stretching exercises, and warming-up. The digit strapping which is often used by climbers is usually a way to reduce stress but has no real effect on protecting the pulleys.

A postoperative rehab has to be followed immediately. It is highly advised to see a hand specialist or physiotherapist. At the beginning, the therapy consists of analgesic and anti-inflammatory medications and stretching techniques.

Once back to climbing practice, the open hand posture is commonly used.

Photo 3.6 This position ensures the best strength distribution in the flexor tendon sheath compared to the crimp grip

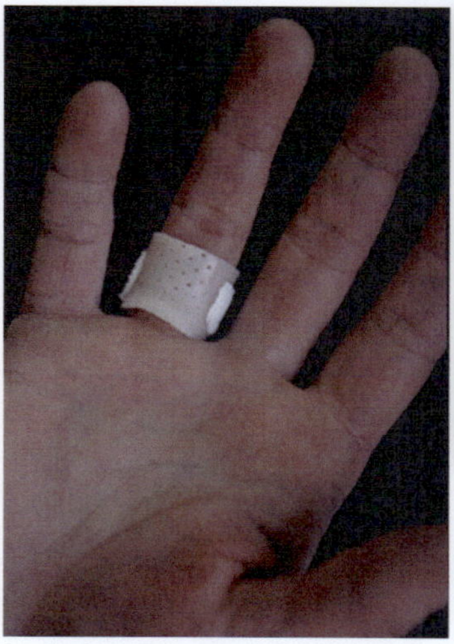

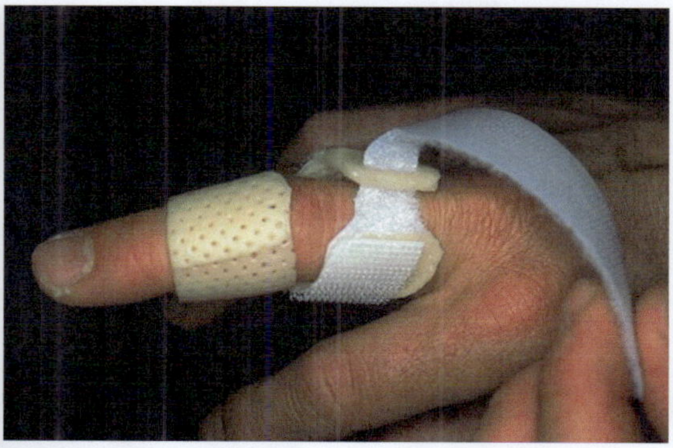

Photos 3.7 and 3.8 Rigid ring protecting the A2 pulley (*left*). Rigid ring protecting the A2 and A4 pulleys (*right*)

3.2.5 Medical Scanning

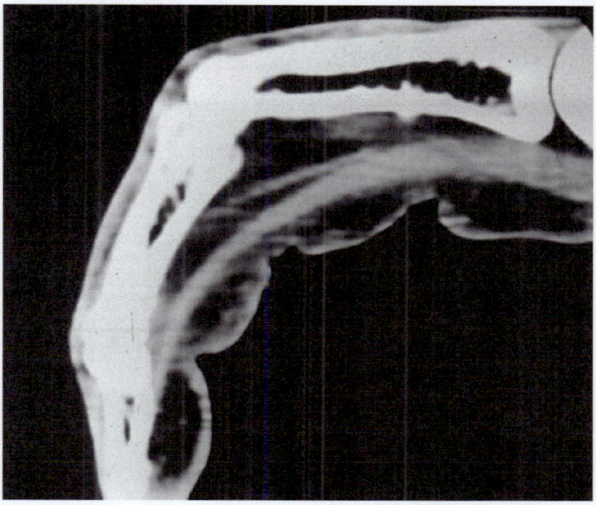

Photo 3.9 Bow stringing due to A2 rupture illustrated with CT scan imaging. Normally, flexor tendons have to be close to the phalanges. In this case, they move forward during the crimp grip

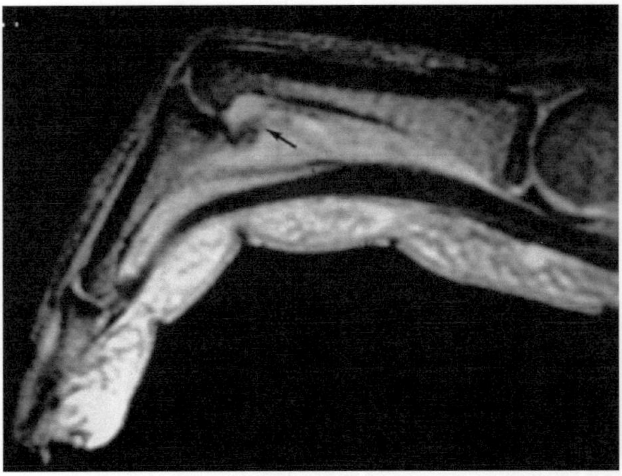

Photo 3.10 MRI illustration of the same injury. Though in black and white, the result is similar to the scan – tendons are represented in black

Pulley Rupture Recap

Symptoms

- A pain is felt when the injured pulley is palpated
- A pain is felt in flexion against resistance from P2 over P1
- A located edema
- A mobility incapacity
- A bow string according to the seriousness of the case

Treatment

- A rigid ring if there's no bow string
- Surgery if there's a bow string

Rest Period

- No activity for 45 days – without any bow string
- No activity for 3 months – surgery

Behavior

- Immediate stopping of climbing.
- Immediate icing and then three times a day.
- Consult a hand specialist.

3.3 Ligament Lesions

3.3.1 Sprains

3.3.1.1 Introduction

Ligaments are structures whose aims are to stabilize laterally any joints. Their function is to maintain the joint surface in contact while in motion.

A sprain is ligamentous distension or a rupture sometimes. Medically speaking, this kind of lesion does not lead to the definitive loss of one's capacities.

The injury may appear right in the ligament area or at the bony junction. Any sprain has to be well-treated for fear of a joint dysfunction.

A sprain can be diagnosed as a minor to a major injury according to the seriousness of the case.

The usual cause of this lesion is a forced movement which leads the finger out of the axis of the hand. Lateral ligaments are more frequently hurt. The three finger joints – MCP, PIP, and DIP – may be affected, but in rock climbing, MPC and DIP are less concerned despite their main role in the practice. MCP joint is mainly used to cling to voluminous holds, thanks to the hyperextension. DIP interferes in crimp grip hyperextending the last phalanx P3 to amplify the contact between the skin and the hold.

The PIP joint is the more often wounded part due to the movements and grips involved in rock climbing. This phenomenon accounts for several reasons:

- PIP is located in a sensitive area, where P1 and P2 phalanges act as a lever. This area is the anatomic and functional center of the finger.
- PIP has only one axis of movement: flexion and extension. Lateral movements are impossible.

Minor sprains are often due to forced lateral movements, whereas major sprains – with volar plate tearing off – occur during forced movements in hyperextension.

In addition to the usual clinical exam, X-rays should be done to visualize a volar plate tearing off – a high-definition view of the lateral side – or a joint fracture which will eventually be treated in surgery.

A sprain is not a minor pathology. An untreated major finger sprain may end up in a joint stiffness and instability. This is why it is very important to pay attention to any swelling or painful bruise on any joint.

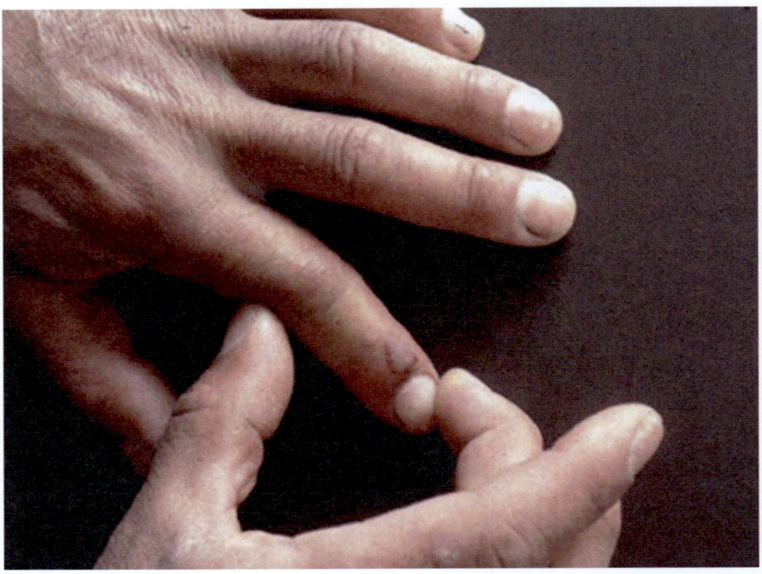

Photo 3.11 Looking for PIP joint instability (index finger). It must be compared with the sound side

Sprain Factors

In rock climbing, several factors are responsible for joint lesions. One of the main causes of sprains is a finger torsion occurring during some of the grips.

Therefore, lesions often involve the lateral ligaments of the PIP joints due to the open hand, one-finger pocket grip commonly used in holes and leaks. Indeed, the movement implies a strong pressure on the finger. Besides, the deepness and the diameter of the hold have a great influence on the risks at stake. According to the diameter of the hold, the number of fingers used may change; one (P3) up to three (P3, P2, P1) phalanges can be necessary to cling to the hold. The use of P3 and P2 may lead to sprains.

The diameter of the hold is another important factor since the finger's mobility is sometimes limited in motion. In some cases, the finger has to pivot inside the hold, so when it's hardly possible, a ligament torsion may happen.

This phenomenon of ligament torsion is also the result of a tough reception on the hold after a dyno or a dynamic movement. This is why it is essential for the climber to have followed a good warm-up and to be in a good state of mind to limit the risks of lesions. These kinds of injuries hardly occur when all the fingers are gathered together to ease the reception on the hold. Indeed, the higher the number of fingers is, the less the lesions are, especially as regards the lateral ligaments. Warm-up is a priority since it is a good way to prepare physically and mentally the body for the next movement. The more recurrent the movements are, the heavier are the shocks undergone by the finger structure and consequently the more frequent are the lesions. Therefore, the number of attempts to get through a route has to be limited especially when the required capacities are higher than the athlete's.

Nonetheless, sprains not only occur in motion but also in falls. More precisely in one-finger grip when the finger may get stuck or be held back. The shock felt by the joints may be hard enough to injure the lateral ligaments or the volar plate. The fingers may also get twisted after a fall on the crash pad or on the ground – in the bouldering category. In that particular case, the sprain is usually palmar and not lateral.

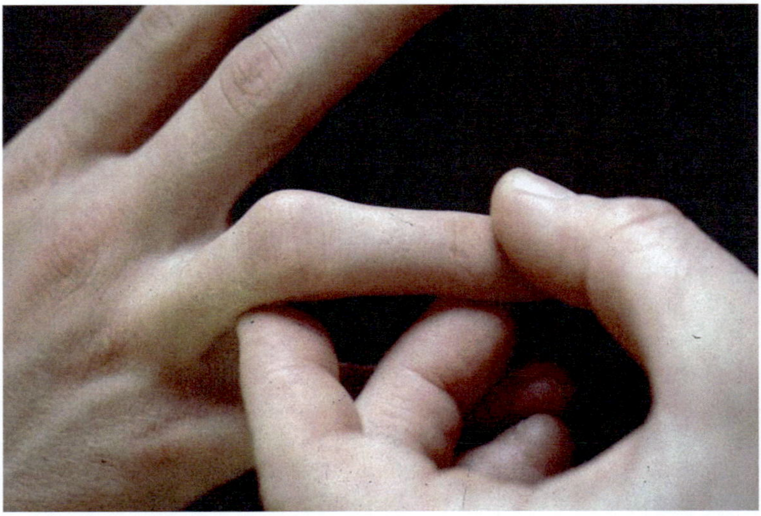

Photo 3.12 The after effects of a neglected radial collateral ligament sprain of the PIP joint (4th finger). The joint is unstable, swollen, and painful

3.3.1.2 Clinical Data

In that kind of trauma, the medical exam is extremely important. The X-ray results generally performed in emergency and showing no fracture may not reveal a potential joint injury. This is why the exam has to be completed with a medical treatment and a consultation with a sports doctor, who will be the only one to make a precise diagnosis. The X-rays will be more convenient to reveal or not a bone tear, which might make the injury much worse.

A sprain is easily detected thanks to the way the sportsman may feel while in motion. Quite often, a failure in mobility and a pain in the finger are present. A bruise may also appear inside the finger in case of a palm sprain. Most of the time, a swelling can also be visible and last for several months, until the ligament repair process is finally over.

The medical checkup is a means to make a precise diagnosis. First, the hand specialist tries to find out any serious pathology which could imply an emergency treatment. Then the joint lesion and the wounded ligament are located. According to the seriousness of the injury – a ligament distension or a ligament tearing off – the final diagnosis is eventually pronounced as a minor sprain or a major one.

The clinical exam is composed of two steps:

- A static phase to check the ligament's state and the joint's stability
- A dynamic phase to make an assessment of the anterior and lateral ligaments

X-rays are used to check if there are any other lesions such as joint fractures or PIP dislocation. X-rays are also a good means to detect any potential bone tear which are quite rare in lateral lesions but present (40 %) in anterior ones.

The seriousness of the injury is classified according to three steps:

- First step: the joint is stable while in motion or in stressful conditions.
- Second step: the joint is stable while in motion, but looseness and hypermobility appear in stressful conditions.
- Third step: the motion phase is abnormal, and instability is obvious in stressful conditions.

The lesion can be lateral, palmar, or dorsal.
As regards the lateral one, three steps define the gravity of the injury:

- First step: for a minor sprain, pain is felt all along the ligaments due to a collateral stretching, but no instability is detected.
- Second step: for a semiminor sprain, a ligament sore is associated with a moderate looseness –inferior to 20 % in the healthy area. The collateral ligament has been torn.
- Third step: for a major sprain, the testing reveals a looseness superior to 20 % in the healthy area. The palmar plate has been injured which leads to an obvious instability.

As regards a former lesion, the palmar plate – whose function is to resist hyperextension – has been hurt. The seriousness of the injury is divided in two phases:

- First phase: no bone fragments are present.
- Second phase: the lateral distal insertions of the palmar plate are torn off.

A sprain may also involve combined lesions especially in torsion motions.

3.3.1.3 Treatment and Behavior

The key thing to do is of course to stop the training practice and ice the finger. A wrong diagnosis or a not full healing might lead to the instability of the entire joint structure. This is why it is highly recommended to consult a hand specialist and do the required X-rays – front and side – as soon as possible.

Meanwhile, the climber will have to ice the finger three times a day until the doctor's consultation, which will define the proper treatment to be followed. Nonetheless, the athlete can tape his finger – knitting it to the next finger or syndactyly. Be careful not to join the fourth finger to the fifth one because the PIP is not at the same level.

Sprains – minor or not – have to be treated seriously because these kinds of lesions may reveal other hidden injuries.

The treatment depends on the seriousness of the sprain and more particularly on the gravity of the lesions and the stability of the injured joint. The therapy aims at regaining or preserving the finger mobility and stability. Any PIP trauma needs a two up to 6 weeks of physiotherapy treatment (active mobilization) to avoid stiffness. The thumb MP is the only one which has to be maintained in full immobility.

Even for a minor sprain, full practice has to be ceased for 3 weeks minimum.

Long Finger MCP Injuries

Minor sprains are treated by active mobilization with the finger knitted to the next one. Therefore, climbing is not allowed for about two up to 3 weeks.

In case of a major sprain, surgery is not always systematic. In fact, surgery may imply several operations such as the stitching of a torn ligament, the replacing of a pulled off bone fragment, or a complementary repair –ligament surgery – for a lesion detected after a while.

PIP Lesions

As regards dorsal injuries and especially minor sprains, the PIP must not be splinted in extension but needs active mobilization for about 2 weeks or up to 3 weeks with a combination of physiotherapist sessions using finger syndactyly.

Concerning palm injuries, due to the retraction of the plaque in the process of healing, a growing stiffness in the PIP in flexion can't be avoided. Therefore, the immediate treatment to be followed – for all the long fingers – consists of an active mobilization.

DIP Lesions

The joint usually remains stable. Forty-five days of full stop is required. If open the injury may infect the tendon system and turn into a phlegmon. In that case, surgery is compulsory, and the athlete will be treated by a complete bathing of the joint and the tendon area. The volar plate never has to be stitched.

Medical Scanning

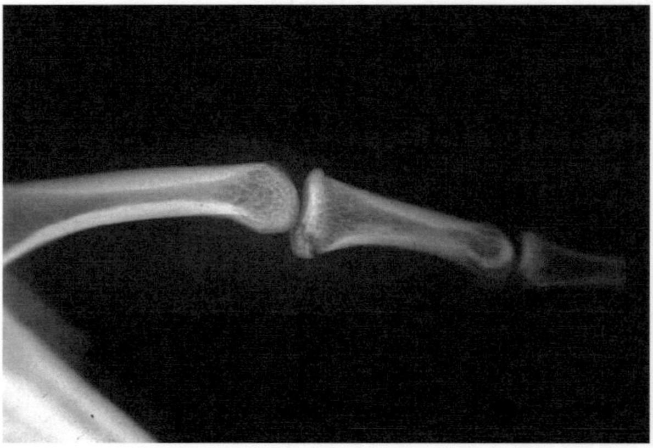

Schema 3.5 X-rays: avulsion of the PIP volar plate (anterior sprain)

Sprain Recap

Symptoms

- A bruise, a swelling, and a finger dysfunction

Treatment

- Immediate active mobilization with or without knitted fingers

Rest Period

- Three weeks minimum

Behavior

- Full stop climbing practice
- Immediate icing and then three times a day
- Consult a doctor

3.3.2 Dislocations

3.3.2.1 Introduction

A dislocation is a displacement or spreading apart of two joint surfaces. The disconnection of these two surfaces often involves a ligament tearing off with or without bone fragments. Joint surfaces, nerves, and blood vessels can be seriously damaged by this kind of lesion. The final risk is of course fragility of the complete joint structure.

This injury can be aggravated by a bone fracture increasing the instability of the joint structure.

This kind of injury hardly happens in climbing. Indeed, when a climber twists his finger in a hole during a grip, he may hurt the collateral ligament rather than the tendon flexors. Nonetheless, a connection can be made between a sprain and a luxation since the latter can be regarded as a sprain which has turned nasty. In climbing, this injury is possible and it is not a minor pathology. This is why this trauma has to be treated and reduced in emergency by a doctor since as time goes by, it gets worse. Indeed, a badly treated luxation leads to a final stiffness and post-traumatic arthritis.

3.3.2.2 Dislocation Factors

Though finger dislocations are quite rare in climbing, risks still remain. Lateral and dorsal lesions affecting the collateral ligaments and the palmar plates essentially have the same causes than sprains. Due to the finger anatomy and biomechanics, some movements are impossible. Therefore, it is obviously easy to imagine what might happen if a finger was stuck into a hold and rotating at the same time.

This injury may occur during torsion movements in the open hand posture in the PIP, the DIP, or the MCP joint area. According to the numbers of fingers used to grip to the hold, the risks to get injured get higher. Indeed, a climber using the one-finger pocket grip to cling to the hold is more likely to face this kind of lesion rather than if he used his four long fingers. One-finger pocket grips are a real strain for the finger joints and more particularly for the PIP. This is why beginner climbers are recommended not to use them, though this injury more often happens during dynamic movements or dynos.

The shock felt by the fingers at the reception on the hold, with the repetition of the movement, undoubtedly accounts for this sort of lesion. The repetitive shocks and the constant weight endured by the same finger structure are bound to hurt the tissues involved to a certain extent.

Though falling is part and parcel of climbing, it is also a factor of dislocations. So the climber may happen to face different kinds of falls and not be able to handle all the parameters which will allow him to be safe. Indeed, most of the time, falls are more or less horizontal – especially when he tries to regain his stability – or more or less violent, according to the height or dynamic actions, Consequently, the first reflex is to reach out his hand in order to protect himself while he is falling. The thumb is the most often wounded limb much more than the other fingers.

During a grip, a finger can get stuck in a hole, the joint capsule can be injured, or even worse a finger bone fragment can be pulled off.

3.3.2.3 Clinical Data

For that kind of lesions, the clinical data are used to determine the circumstances of the incident. The key thing is to find out whether the luxation is associated or not with a bone fracture. X-rays are a priority to make a proper assessment of the joint's state. A difference has to be made between the stiffness involving a healthy joint and the stiffness involving a wounded joint, the latter requiring surgery before regaining mobility.

Lateral injuries can be classified in three steps:

- First step: the collateral ligament is extended. The joint is stable.
- Second step: the collateral ligament is torn off. A moderate looseness of the joint can be noticed.
- Third step: the lesion has affected the volar plate. A chronic instability is visible.

As regards palm injuries – which affect the beginning of the volar plate, the structure opposing hyperextension – two phases are possible:

- First phase: no bone fragment is present. The lesion is painful but does not affect the stability.
- Second phase: the distal lateral insertions of the volar plate are torn off.

3.3.2.4 Treatment and Behavior

The first thing to do is of course to stop climbing. Whatever the injury, the climber is unable to carry on the ongoing effort. A luxation can seriously damage the hand structure if not treated properly. This is why it is important to be extremely cautious before going immediately to the emergency and have some X-rays done before consulting a hand specialist.

Quite often, the patient delays the consultation and the finger has already got stiff and deformed, which makes the treatment very difficult. This is why an early and precise diagnosis is the priority. The aim then is to regain mobility and to reduce the stiffness of the finger.

Most of the time, stiffness is due to a too long or a not appropriate immobilization after a joint trauma. It can also be the result of a too late or an inadequate physiotherapist prescription. The problem is that stiffness can spread over the other joints including those of the next finger.

A post-surgery physiotherapy is the key to reduce stiffness. Physiotherapy combined with anti-inflammatory treatment is systematic to reduce joint edema which prevents mobility.

The treatment of course varies according to the seriousness of the luxation and especially in case of complete ligament tearing off or not.

MCP Joint Injuries

These injuries are difficult to be diagnosed. The clinical symptoms can be an axial rotation of the finger combined with a lateral curvature. These symptoms reveal the blocking of P1's head and require surgery. It is a priority not to pull at the finger. After surgery, the MCP is partial immobilized in a flex position with splint, which enables the use of PIP and light MCP mobilization. The immobilization has to last from three to 6 weeks.

When the climber goes back to practice, he has to train with joined fingers to protect the joints.

Palmar dislocations of the MCP are extremely rare, contrary to the dorsal ones which are more common.

PIP Joint Injuries

The dislocations of this joint can be either on the palmar area – though quite rare – or on the lateral or posterior one.

In the palmar area, the phalanx's head which passes through the lateral strips of the extensor may also tear off the central strips of the extensor. When bone fragments are present, surgery is required. In case of the absence of any bone fragments, the doctor can check the healthiness of the central strip, by flexing and extending completely the finger under a certain pressure. If everything is okay, an orthopedic treatment with the fingers maintained fixed will be possible. Regular X-rays will complete this treatment.

In case of breaking of the central strip of the extensor, surgery is indispensable. This kind of lesion can be tested. Surgery is also required when the patient is not able to maintain a complete extension of the PIP against resistance or when a bone fragment is present.

During healing time, the palmar plate tends to shrink, which leads to a stiffness in the PIP especially in flexion.

As regards lateral lesions, the doctor makes sure that no instability is present while flexing and extending. The treatment is identical to serious sprain. The finger is often swollen and a certain amount of stiffness is commonly present.

Obviously, the treatment depends on the seriousness of the injuries and consequently on the degree of stability. For a moderate looseness, the finger will be mobilized for 3 weeks. In case of instability with a looseness superior to 20°, a ligament surgery will be required.

Dorsal lesions result from a brutal flexion against resistance of the PIP, and this kind of luxation is quite frequent but is due to a lateral trauma combined with torsion.

DIP Joint Injuries

The symptoms are identical to those of the PIP joint, but the risks to get an open fracture – exposing the tendon flexor or extensor – are higher. In that precise case, surgery is compulsory.

To conclude, all these kinds of lesions have to be well diagnosed for fear of seriously damaging the finger functions.

Medical Scanning

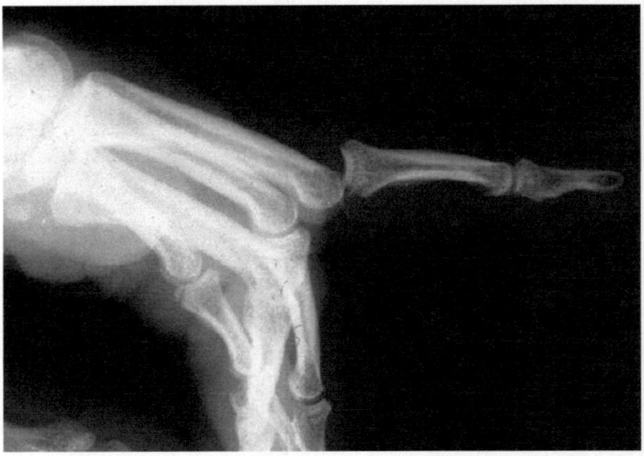

Schema 3.6 Typical dorsal PIP joint dislocation

Luxation Recap

Symptoms

- A bruise, a swelling, limited mobility of the finger, and joint instability

Treatment

- Joined fingers – fixed to the next one – or surgery for serious injuries, in case of bone fragments for instance

Rest Period

- About 6 weeks according to the seriousness of the injury

Behavior

- Climbing has to be stopped immediately.
- Immediate icing and then three times a week.
- Consult a hand specialist right away.
- Make X-rays.

3.4 Bone Injuries

3.4.1 Introduction

In trauma, a fracture is when a bone breaks off.

Therefore, from the anatomy point of view, the lesions can be located in the carpals, the metacarpals, or the phalanges. According to the localization, the treatment differs but the aim is the same, that is, regaining a full functionality of the hand with no stiffness or pain and gradually train the hand to seize objects with the palm and eventually with the fingers. As regards climbing, the aim is to be able to seize all the possible holds such as buckets, large holds, or small edges.

A lot of fractures first considered as minor can unfortunately be badly treated and lead to a stiffness in one or several fingers.

These pathologies will be treated according to their localizations.

3.4.2 Clinical Data

Several symptoms indicate this kind of lesion. The athlete may feel a sudden and localized pain. Most of the time, it occurs after a shock or a fall. Because of this lesion, some movements are impossible due to the pain felt. Quite often, a swelling is visible in the wounded area. A distortion may also appear in case of a bone displacement.

3.4.3 *Fracture Factors*

In climbing, these kinds of pathologies hardly happen, and most of the time, they are due to other factors independent to the climbing-specific movements. Exterior factors such as a fall with one or several fingers stuck in a hold or a failed reception outside the crash pads account for these injuries.

3.4.4 *Treatment*

3.4.4.1 Phalangeal Fractures

The main issue in this kind of injury is how to regain a full mobility of the wounded part without any finger stiffness. Complete immobilization is to be banned. In case of a closed fracture, with no bone displacement or instability or joint problems after reduction, the finger joined to the next one is a good means to ease the repair and to maintain the joint mobility at the same time. But, on the contrary, if it is an open fracture, with a bone displacement, instability, or a joint problem, surgery is required. The principal aim is of course to regain a complete joint mobility as soon as possible, with no stiffness or grip. The treatment is always functional or surgical and then followed up by 2 weeks' time with joined fingers.

Concerning a fracture of the P3, the finger is immobilized with a splint for 6 weeks, allowing the PIP joint to be mobile.

As regards the P1 phalanx, about 5–8 weeks is necessary for the bones to knit together, which implies a full stop in practice of 8 weeks minimum. The follow-up reeducation sessions are vital.

Concerning the P2 phalanx, the bone knitting occurs after 8 up to 10 weeks, which means that full climbing practice has to be stopped for 10 weeks minimum.

The bone knitting is followed up by regular rays, which are checked by the doctor who decides whether or not the athlete is fit to go back to climbing practice.

3.4.4.2 Metacarpal Fractures

Treatment is the same to the phalange one. It is always functional or surgical and then followed up by a 2 weeks' time with joined fingers.

3.4.4.3 Carpal Bone Fractures

The scaphoid fracture is the most current as regards carpal bone fractures. In rock climbing, this injury results from a fall on the palm during bouldering activities for instance. But actually, this pathology is quite rare.

At landing due to the hyperextension of the wrist, the scaphoid happens to be stuck, which leads to breaking.

The clinical signs of this lesion are visible around the wrist area, at the back of the thumb near the wrist with pain during palpation or in motion. This injury may be invisible on X-rays. Vivid pains are not always combined with the symptoms, which may result in delayed diagnoses or misjudgments. New X-rays made 10 days later can bring new elements to define the origin of the pain.

Concerning most of the scaphoid fractures with stable elements and no displacement, a 90 days' immobilization is necessary to assure bone knitting.

As for the instable fractures or osteoarthritis due to a too late consultation, surgery is required.

3.4.5 Scanning Images

3.4.5.1 Fractures of Two Phalanges P1

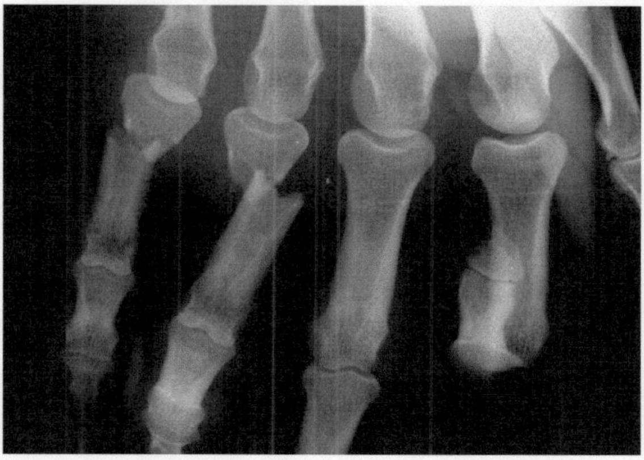

Schema 3.7 Unstable fractures of two phalangeal P1

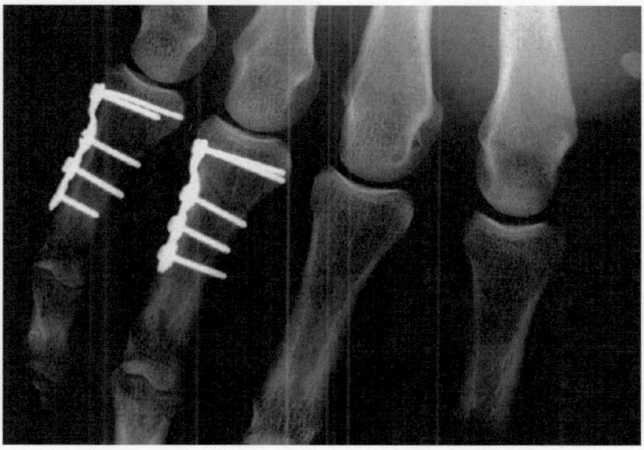

Schema 3.8 Two unstable fractures of phalangeal P1 treated with ORIF (rigid internal fixation)

3.4.6 Scaphoid Fracture

Schema 3.9 Scaphoid fracture

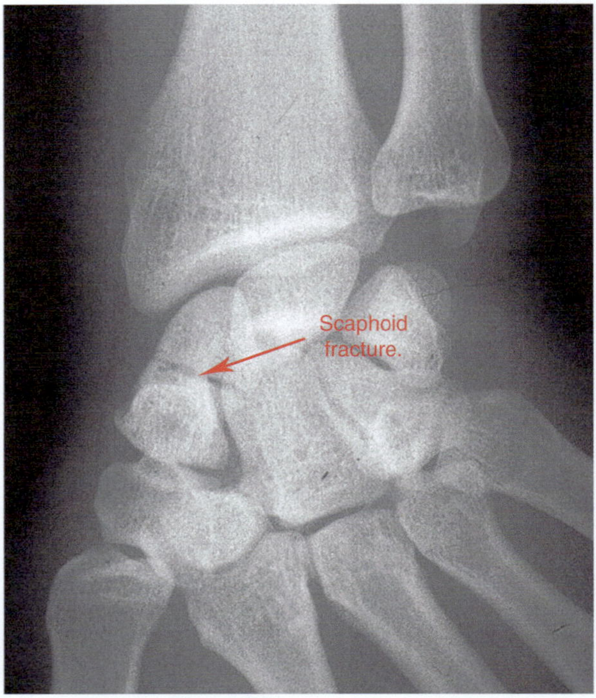

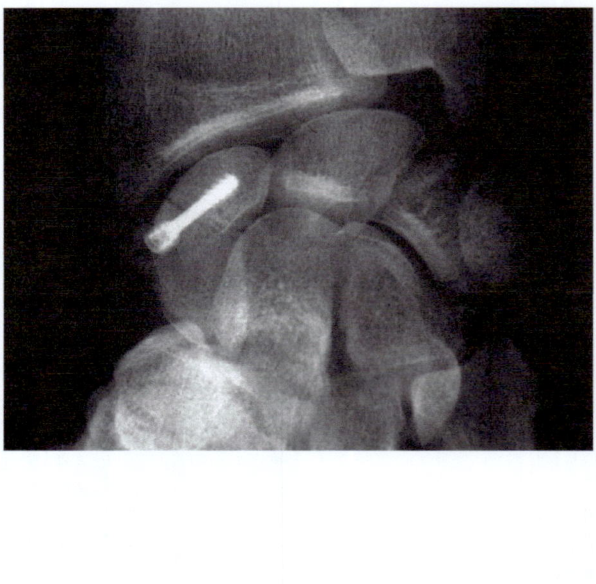

Schema 3.10 Scaphoid fracture treated with ORIF (screw)

Fracture Recap

Symptoms

- Pain during palpation and in motion, finger functional disorder according to the area

Treatment

- Immediate active mobilization for a non-open, stable, and non-joint fracture with no difficulty in finger mobility.
- Surgery is required for all the other cases.

Rest Period

- 8 weeks for a P1 fracture
- 10 weeks for a P2 fracture
- 12 weeks for a P3 fracture
- 8 weeks for metacarpals
- 3 months for carpal bones

Behavior

- Full stop of climbing practice.
- Immediate icing is required and then three times a day.
- Consult a doctor.

3.5 Muscle Injuries

3.5.1 Lumbrical Muscle Tearing Off

3.5.1.1 Introduction

This kind of lesion is extremely rare in rock climbing. For the moment, few of these cases have been reported. This lesion usually concerns the third or fourth finger in open hand posture. Indeed, the small lumbrical muscles which are extended between the flexor and extensor tendons are extremely strained in this position of the hand, and they may not be able to endure a too heavy weight.

The climbers who previously suffered from this kind of lesion were able to use the one-finger pocket grip two months after their return to practice. The sensation of the grip was different, and they did not manage to regain the former level.

Photo 3.13 One-digit grip

3.5.1.2 Lumbrical Muscle Tearing Off Factors

One-finger pocket grip seems to be the main factor of this kind of injury. When the finger is straightened and the others are flexed to gain as much strength as possible, a tearing of this little muscle is possible. Of course, other factors such as a lack of warm-up and a bad hydration enhance the eventuality of such a lesion.

Schema 3.11 Lumbrical symptoms (Illustration inspired by Schweizer (A.), *Lumbrical Tears in Rock Climbers*, The Journal of Hand Surgery, 2002)

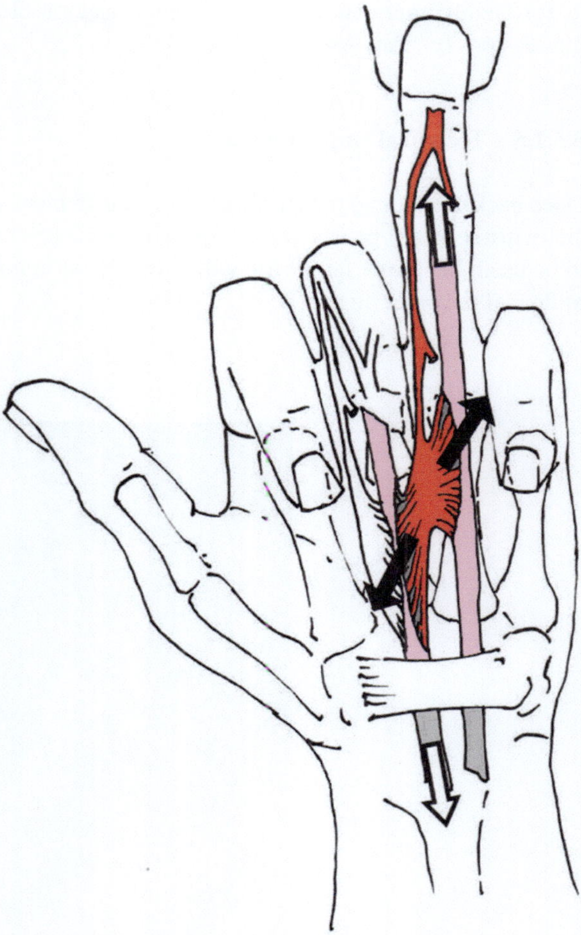

3.5.1.3 Clinical Data

Pains are felt in the palm of the hand area. Palpation can stand out this vivid pain. The pain is quite sudden and the climber may hear a cracking sound. An edema or a bruise may also appear. The clinical exam can be completed by a scanning.

3.5.1.4 Treatment and Behavior

The climber has to cease any practice immediately and ice his hand three times a day for a week. According to the seriousness of the injury, a period of rest from 10 days to a month is required. The best thing to do is to join the injured finger to the next one. Once back to practice, the climber has to avoid any one-finger pocket grip for about two months. During the first climbing sessions, he can have his finger taped to the next one in order to relieve the possible tensions.

It is highly important to stretch the lumbrical muscles correctly after each climbing sessions (cf. Part Assessment and Safety).

3.5.1.5 Technical Improvement

Once back to training, the climber has to change his way of seizing the holds, especially in one-finger pocket grips. Most of the time, the two-finger positions will have to be used. In case of failure, he will have to bear in mind that he won't be allowed to flex all the other fingers.

Photo 3.14 One-digit grip (outdoor)

Lumbrical Tearing Off Recap

Symptoms

- Bruise, edema, pain in the palm of the hand

Treatment

- The finger is joined to the next one.

Rest Period

- Immobilization for 3 weeks to one month according to the seriousness of the injury.

Treatment

- Full stop in climbing.
- Immediate icing and then three times a day.
- Consult a specialist right away.
- Make a scanning.

3.6 Other Injuries

3.6.1 Ring Finger

For climbers' safety and to avoid some extremely dangerous practices, it is important to deal with that part of recommendations.

Indeed, wearing rings increases the risks of finger pulling out. Rings may get stuck in a hold and consequently lead to a finger tearing off. The lesions caused by rings – often quite tricky – may be difficult to repair surgically. The ring finger lesion can be compared to taking off a glove. In fact, the skin slides off the finger until complete tearing which leads to bone trauma – common fractures in the PIP area – and blood vessel trauma.

Therefore, climbing with a ring is to be banned even though this kind of injury remains quite rare. It is also very risky to put a finger in a hole or an auto-blocking belay device in order to be supported. In case of a fall, the consequences are the same than those of wearing a ring, that is to say, the finger gets stuck and eventually gets torn.

Another lesion with similar consequences to that of the ring finger is when the rope folds around a finger. In case of a fall, the climber may be tempted to seize the rope that might then twist around a finger. After, when the rope is suddenly extended because of the climber's weight in his harness, the finger is squeezed and may be torn off.

Other cases resulting in a finger tearing off, such as trying to catch a quick draw during a fall, have been reported.

Therefore, it is highly important to bear in mind that though these lesions barely occur, the risks nonetheless remain high and the consequences may be extremely damaging. Consequently, it is wiser to follow the safety recommendations to avoid any problems in climbing.

3.6.1.1 Behavior

In case of tearing, cutting, or crashing of a finger, here are the safety gestures to apply:

1. Apply a clean compress on the injured area and maintain it until the bleeding stops.
2. Make a proper bandage up to the wrist and keep the lesion tight.
3. Making a tourniquet is banned and might be damaging for the follow-up treatment.
4. The cut finger has to be picked up, stripped off dirt, cleaned with water without any brushing, wrapped up in a piece of fabric, and finally stored away in a plastic bag. It is much better to put this bag in another one containing ice. It is highly recommended to turn the bag every 10 min to change the finger's position.
5. The injured climber must not drink anything until the induction of anesthesia.
6. The mobile emergency medical service has to be called immediately.

The lapse of time between the incident and the surgery must not be over 6 h. A repair surgery can be made according to the staff decision.

Even if the repair surgery is successful, a certain physical, sensitive, and functional capacity will be lost.

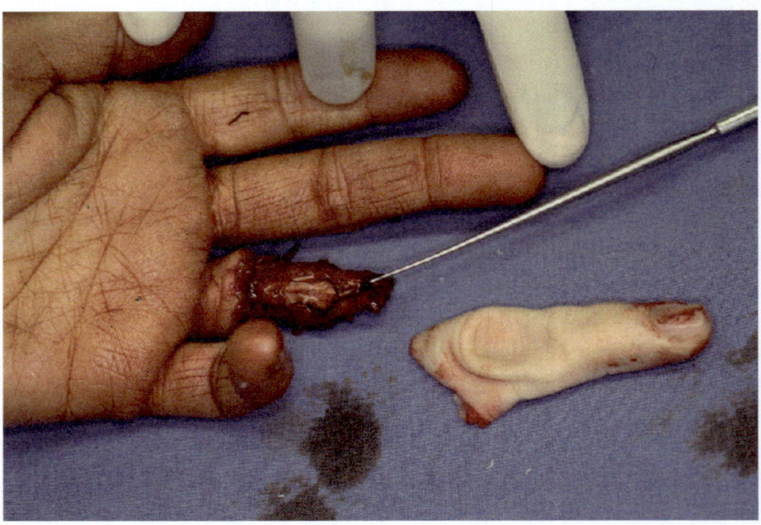

Photo 3.15 Ring finger (4th digit)

3.6.2 Trigger Finger

3.6.2.1 Introduction

The trigger finger is a current pathology. It is when a tendon happens to get stuck from time to time during a finger flexion. Usually, the tendon is supposed to remain flat along the bone fingers, thanks to the pulleys. The tendon is enveloped by a synovial sheath making the sliding easier. When the tendon is a bit thicker in an area, the sliding is slightly impeded and the A1 pulley is stuck. This struggle in flexion is combined with a start in extension and the finger extends suddenly.

A pain is felt in the palm area at the base of the finger. Most of the time, the wounded athlete feels as if the finger was gripping in the PIP joint, but this location – pointed by the athlete – is actually wrong.

3.6.2.2 Trigger Finger Factors

One of the possible causes of this problem is the inflammation of the synovial sheath. Gradually, it becomes a mechanical obstacle for the finger's mobilization.

The repetition of the same movement may also be another factor of this pathology.

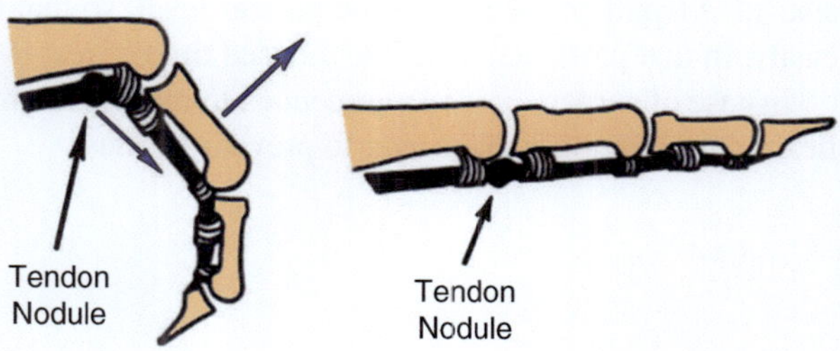

Schema 3.12 The trigger finger (Illustration inspired by Danowski (RG), Chanussot (JC), *Traumatologie du sport*, 2005, Masson)

3.6.2.3 Clinical Data

The diagnosis relies essentially on the clinical exam. The climber feels as if his finger was griping while he is flexing or extending one of them.

At a more serious phase, the climber has to use his second hand to unblock the finger still in flexion.

No other side exams are necessary. A scan or an MRI can confirm the diagnosis showing an infiltration in the synovial sheath or a tendon inflammation.

3.6.2.4 Treatment and Behavior

The treatment depends on the evolution of the trauma. In case of a minor pathology, a full stop in climbing practice and a period of rest for the injured finger may help and ensure a proper recovery.

A corticoid infiltration – an anti-inflammatory treatment – to reduce the swelling of the tendon may be required. Quite often, a second infiltration is necessary to improve the results. This treatment is efficient but quite temporary. The infiltrations can be repeated two or three times but must not be overused. Indeed, cortisone may make the tendon fragile and break it. The rest period is about 10 days minimum, if the treatment has been efficient and the nodule has disappeared. As long as the swelling is present, any climbing practice is prohibited.

Surgery can be practiced when the former treatment has been inefficient. It consists of A1 pulley opening, at the palmar level, so that the tendon may slide much easily. In that particular case, the advised time of rest is about 3 weeks minimum.

In case of surgery, it is recommended to make an immediate auto-reeducation by flexing and extending the finger to prevent stiffness.

Photo 3.16

Chapter 4
Surgical Techniques

4.1 Tendon Lesions

Regarding climbers' pathology, spontaneous isolated tendon rupture doesn't often occur. Closed flexor tendon avulsion at the distal insertion level is much more a rugby player lesion (jersey finger). Actually climbing reinforces osteo-tendinous attachment to the distal phalanx (P3). If rupture occurs, the treatment doesn't differ that of any flexor tendon injury, and in this case, distal insertion is provided.

S. Gnecchi, F. Moutet, *Hand and Finger Injuries in Rock Climbers*,
Sports and Traumatology, DOI 10.1007/978-3-319-16790-9_4

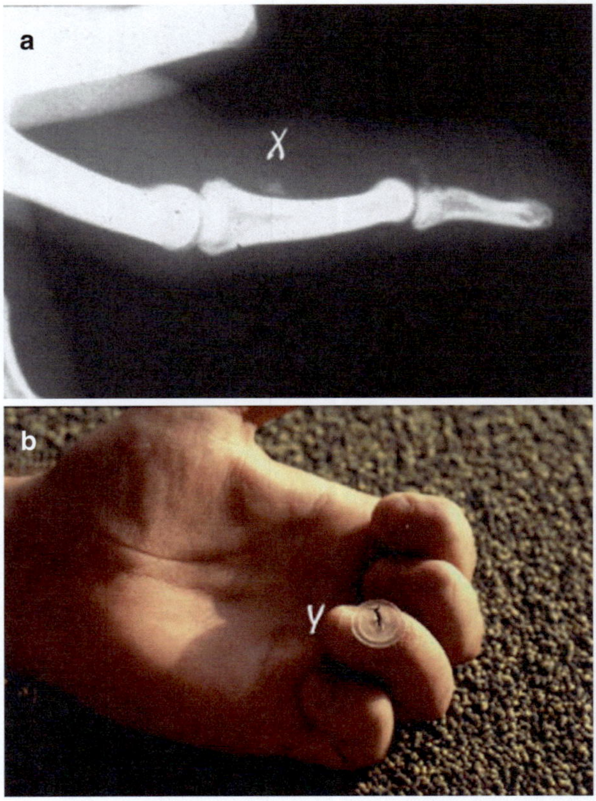

Schema 4.1 (**a**, **b**) Jersey finger. The avulsed flexor digitorum profundus (FDP) is pulled up in the digital sheath (X). It is fixed back in place through the bone of the distal phalanx (P3). The suture is protected by a button fixed over the nail that will be pulled out on the 45th day postop (Y). Active mobilization, with no resistance at all, is allowed

On the same way, any section occurring during climbing practice will be treated by tendon suture, early active motion with a protection splint, and rehabilitation. The suture must be strong enough to allow mobilization, thin enough to allow a perfect gliding of the repaired tendon in its sheath, and in an unloading position to avoid strong tension. In such lesions, the key point is – whatever the zone is – not to apply too much strength too early on the repaired digit at least for 2 months.

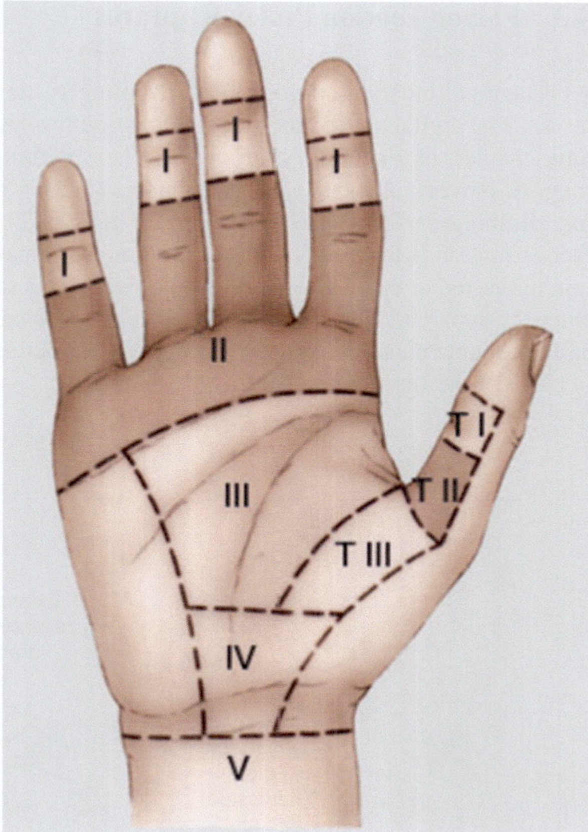

Schema 4.2 Flexor tendon zones international classification (Verdan & Michon). Zone II is the ancient "no man's land." It became the "one man land." The one is and must be the hand surgeon

4.2 Ganglion Cysts

At the wrist level and at the digitopalmar crease (tenosynovial cyst), they are not more frequent as regards the general population. Usually they are not due to a traumatic lesion but rather to overuse conditions. They may be easily cured by surgical resection. People have to bear in mind that this phenomenon is an inflammatory response of the synovial sheath or conjunctive tissue transformation; consequently, cysts have to be removed only if they are still present after a period of 6 months on. That is especially true at the wrist level where they can come and go several times before stabilization.

4.3 Flexor Tendon Pulley Ruptures

As regards climbers' activities, pulley rupture is the most common upper limb lesion. The digital sheath and flexor tendon pulley system, as shown in the anatomy chapter, is a necessary structure. When partially or totally destroyed, the range of movement and the grip strength are severely impaired especially regarding climbing performance. So especially at A2 (75 %) and A4 level, surgical reconstruction is necessary. Reconstruction uses a part of the extensor retinaculum harvested at the dorsal aspect of the wrist. This structure is the pulley of the extensor apparatus and then has the same mechanical properties and histologically the same nature, so it's the perfect reconstruction material. The graft is put in place of the ruptured pulley.

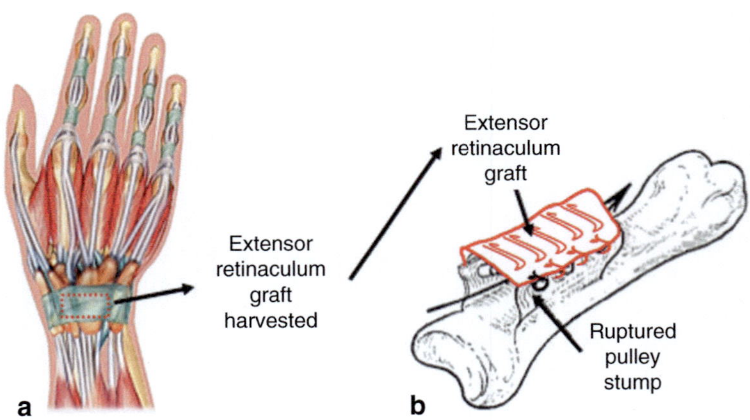

Schema 4.3 Digital pulley reconstruction. (**a**) A retinaculum extensor graft is harvested at the dorsal aspect of the wrist. (**b**) The graft is fixed to the stumps of the ancient pulley

That procedure requires a 3-month total rest period: a month and a half with a massive splint (wrist flexed at 30°, MP joints flexed at 70/80°) and then another month and a half with a rigid ring which protects the reconstructed pulley. Return to climbing may be authorized only after that "incompressible" 3-month period. The climber has to respect this complete rest, because if he doesn't, a second rupture will occur.

When a partial rupture occurs, surgery is not always necessary. In such cases, wearing a rigid ring protecting the concerned pulley, for a month and a half, will be usually sufficient.

Anyway, one must emphasize the imperative conditions of return to climb: hydration, muscle stretching exercises, warm-up, and progressive difficulty upgrading.

4.4 Sprains and Dislocations of the Long Fingers

Sprains are part and parcel of any sports including climbing. Therefore, treatment is quite always conservative. That is to say treatment doesn't usually require surgery, and an immediate active mobilization with a syndactyly splint is quite enough.

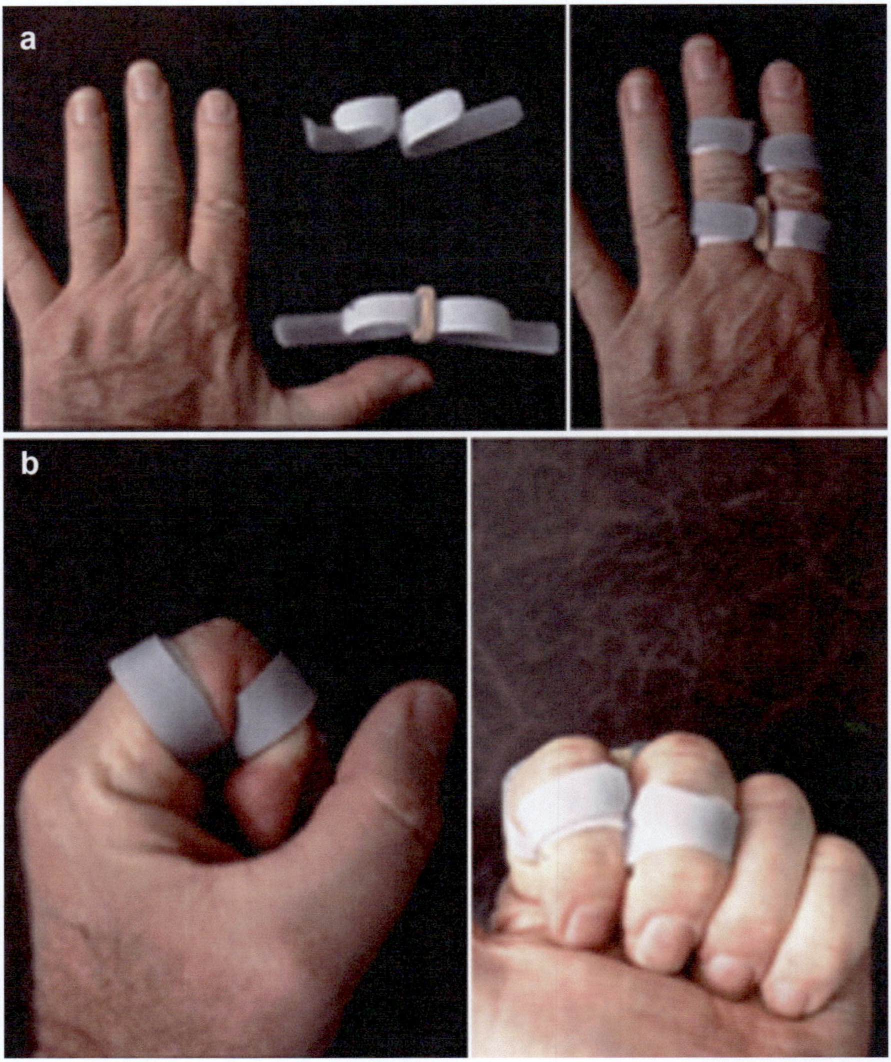

Schema 4.4 Sprain and dislocation treatment: syndactyly. (**a**) The splint uses the clear index finger (index) to correct the long finger axis (sprain at the PIP joint level). (**b**) The splint allows full flexion to insure immediate active mobilization

But one must be aware of the fact that any injuries – sprains or dislocations – require an X-ray examination to look for a bony fragment avulsion. If the presence of a fragment located 20 % over the articular surface of the concerned joint is detected, surgery may be necessary. Only a hand surgeon can perform this specific operation, followed of course by a specialized and continuous rehabilitation for at least a 2-month period.

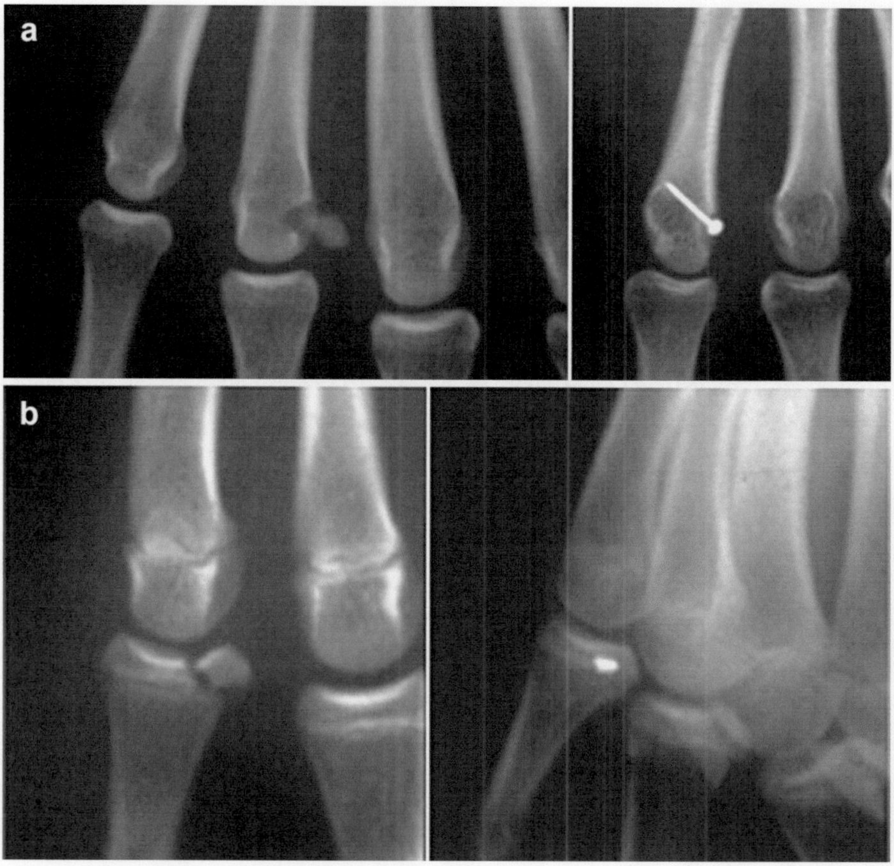

Schema 4.5 Sprains with bony fragment. (**a**) Avulsion and fixation of a bony fragment at RCL of the 4th finger MCP joint (peg). (**b**) Avulsion and fixation of a bony fragment at RCL of the 5th finger MCP joint (screw)

Regarding dislocations of the long finger joints, after reduction and control X-ray examination, the treatment is exactly the same: immediate active mobilization is always better than early active mobilization, using syndactyly if needed.

4.5 Sprains and Dislocations of the Thumb

Regarding the thumb, whatever the lesion is – sprain or dislocation – the aim is to obtain a strong, painless, and stable pinch between the thumb and the other fingers. If an abnormal laxity is discovered while testing the metacarpophalangeal joint (MCP) in light flexion (30/40°) at the ulnar collateral ligament (UCL) or radial collateral ligament (RCL) as well, the ruptured ligament has to be surgically repaired. If it is not painful, a thumb weakness will occur with a severe pinch grasp dysfunction. As for all the other joints, an X-ray examination is always needed. Severe sprains with any avulsed bony fragments lead to a surgical treatment (Table I). Anyway, all these lesions need to be immobilized in a rigid splint or in a cast for 1 month.

While MCP of the thumb is a very well-known lesion, dislocation may sometimes be misunderstood. The proximal phalanx (P1) is propelled on the first metacarpal dorsal aspect. If you just pull straightly to reduce the dislocation, you will block the sesamoid bones into the MCP joint space. You have to use the Farabeuf's maneuver which consists of pushing the dislocated phalanx still erected on the dorsum of the first metacarpal bone and, after MCP joint space perception, flex the MCP joint.

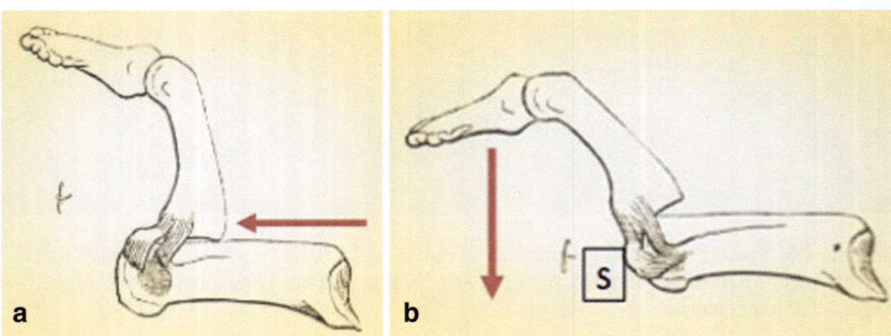

Schema 4.6 Farabeuf's maneuver for MCP joint of the thumb dislocation. (**a**) The proximal phalanx is pushed on the back of the first metacarpal bone "like on a table." (**b**) Then the MCP joint is flexed avoiding sesamoid bone (S) incarceration between the MCP joint articular surfaces

After reduction and an X-ray examination, looking for a UCL or RCL rupture is highly recommended. Actually, a ligament rupture reinforces the diagnosis of a severe sprain injury and requires surgery to restore stability.

4.6 Metacarpal and Phalangeal Fractures

As mentioned in a former chapter, X-ray examinations are necessary to detect fracture. Three types of treatment are possible:

- *Conservative treatment*: an immediate active mobilization when the fracture is closed, stable, nonarticular, and without any digital flexion harmony disruption.
- *Surgical treatment*: an osseous rigid internal fixation (ORIF) is used to allow early active mobilization in order to quickly restore hand function, after a 1–2-month rehabilitation period. That's the reason why it has to be precise and resistant.

This treatment is indicated when the fracture is open and/or unstable/articular and/or leads to a digital flexion disharmony.

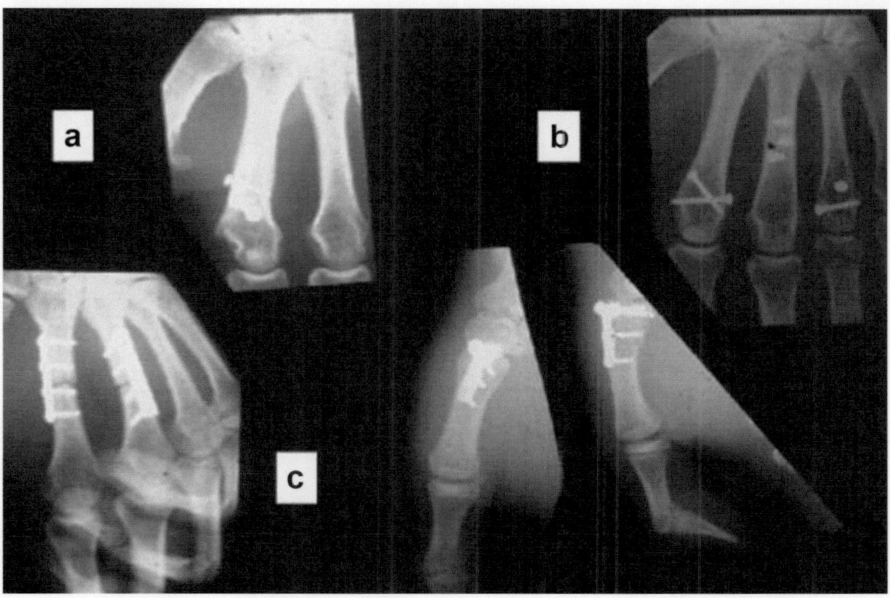

Schema 4.7 Metacarpal fractures (ORIF). (**a**) Blades; straight, L or T shaped. (**b**) Isolated screws

- *Orthopedic classical immobilization*: this treatment is never indicated for climbers and for this type of fractures because it leads to stiffness and poor results.

4.7 Carpal Bones Fractures

Regarding carpal bones, one has to bear in mind that they are all articular bones and have to be carefully examined on the X-ray before choosing the type of treatment. The scaphoid fracture is the most frequent. It requires surgery as soon as displacement jeopardizes the normal wrist anatomy, and then ORIF is applied.

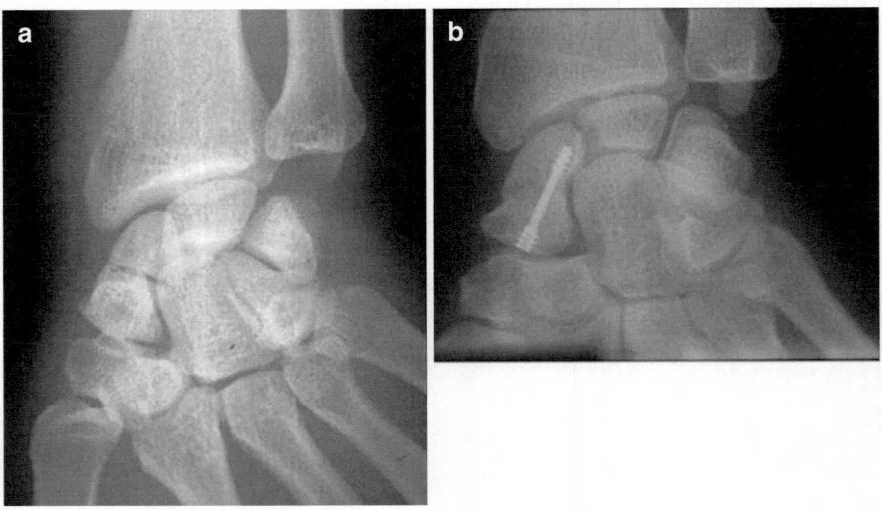

Schema 4.8 Scaphoid fracture (ORIF). (**a**) Displaced corporeal scaphoid fracture. (**b**) ORIF with an intramedullary screw

If it is not, a plaster cast is indicated for not less than 3 months to obtain a complete consolidation. Isolated fractures of the other carpal bones are usually immobilized with a plaster cast for 45 days.

Complex articular wrist lesions, open or not, such as dislocations, fracture-dislocations, or intracarpal instability, follow severe injuries and combine fractures and severe ligament ruptures. These lesions always need surgical reduction, stabilization, and repair. The function restoration needs over a 3-month rehabilitation period and a 6-month follow-up for a global, often impaired, function recovering,

4.8 Ring Finger

It is a very severe lesion which leads to a partial or total skin and neurovascular bundle avulsion. If amputation occurs, replantation is not always possible. Nonetheless, whatever the gravity of the lesion is, the ring finger always represents a severe contusion of the digit and must be considered like it. The treatment will always imply a long and difficult rehabilitation for functional recuperation. Therefore, the main piece of advice to be given is never to wear a ring while climbing. If the ring happened to be stuck anywhere at any time, it could strip a whole finger (see photo 3.5 in chapter anatomy).

4.9 Trigger Finger

Triggering is not a very frequent occurrence in climbers. It can appear when the flexor tendon rubs too heavily under the A1 pulley. Then the thickened flexor tendon envelop may be blocked under A1 in flexion, and extensor tendon is not strong enough to extend the finger. Surgery will cure it by A1 pulley opening. After healing (2–3 weeks), the function is usually totally restored.

4.10 Remarque

All surgery regarding climbers' hand trauma is a 1-day surgery (except for replantation) under peripheral anesthesia. Actually, the main problem is not the surgery itself but the importance for the patient to understand how primordial it is to respect the duration of rehabilitation and more specifically that of the rest period.

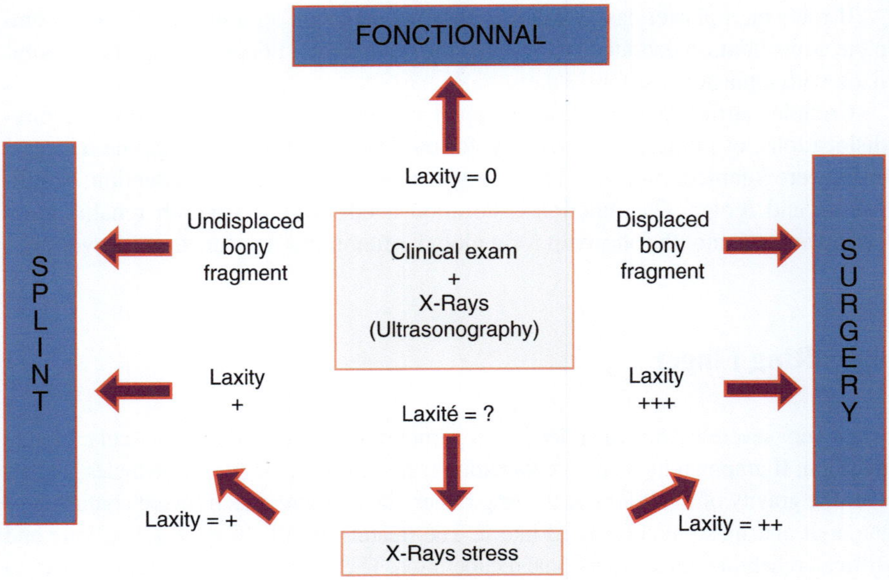

Schema 4.9 Table I. Sprains of the MCP joint of the thumb algorithm

Photo 4.1

Chapter 5
Training and Back to Training

Photo 5.1

S. Gnecchi, F. Moutet, *Hand and Finger Injuries in Rock Climbers*,
Sports and Traumatology, DOI 10.1007/978-3-319-16790-9_5

5.1 Climbing Different Grips

5.1.1 Strains of the Various Grips

The position of a finger can be either straight, arched, or hooked. The hand morphology, the various grips available outdoors or indoors (CG), the climber's level, and his personal abilities are factors which determine the possibilities of seizing a hold.

The fresh climber prefers using the crimp grip. A maximum strength is liberated while gripping, and in a way, this fact reassures the athlete – emotional pole. As he gets more experience, learning new movements, his former crimp grip posture turns gradually into the open hand one. These various changes depend on his physical and emotional skills but also on his own sensations. It is important to develop all the possible tactics by training on various supports – different degrees of steepness, orientations, sizes, spaces, numbers of holds – especially outdoors in order to face any situations.

Reducing the seizing of holds to one posture – crimp grip or open hand grip – would be like decreasing the numerous possibilities available in rock climbing. In addition to finger locks and fist jams, some holds can be tackled combining the crimp and open hand grip using all the fingers in turn. These grips are essentially present outdoors, but some factors make them compulsory in CG. The steepness of a hold can require what will be called a mixed posture, that is to say, a crimp grip for some fingers and an open hand posture for others. A more developed musculature is consequently required between the deep finger flexors (DCF) and the superficial finger flexors (SCF). The finger common extensors (FCE) play a main part (cf. ch. Assessment and Safety). And later in the book, it will be demonstrated how important it is to reinforce them, to improve the stability of the joint system, and to support the strain endured, especially in the open hand posture.

To conclude, it is rare to find a hold perfectly linear as regards grips, especially in natural sites. The angles, width, depth, and steepness of the hold are various factors which give a vast number of actions to each finger. Therefore, according to these factors, the surface in contact or the necessary strength will change.

Hand morphology is also an important matter. Bearing in mind the differences between the four fingers and comparing the climbers' hands, it becomes quite understandable why some of them can't always use the open hand posture or the crimping one.

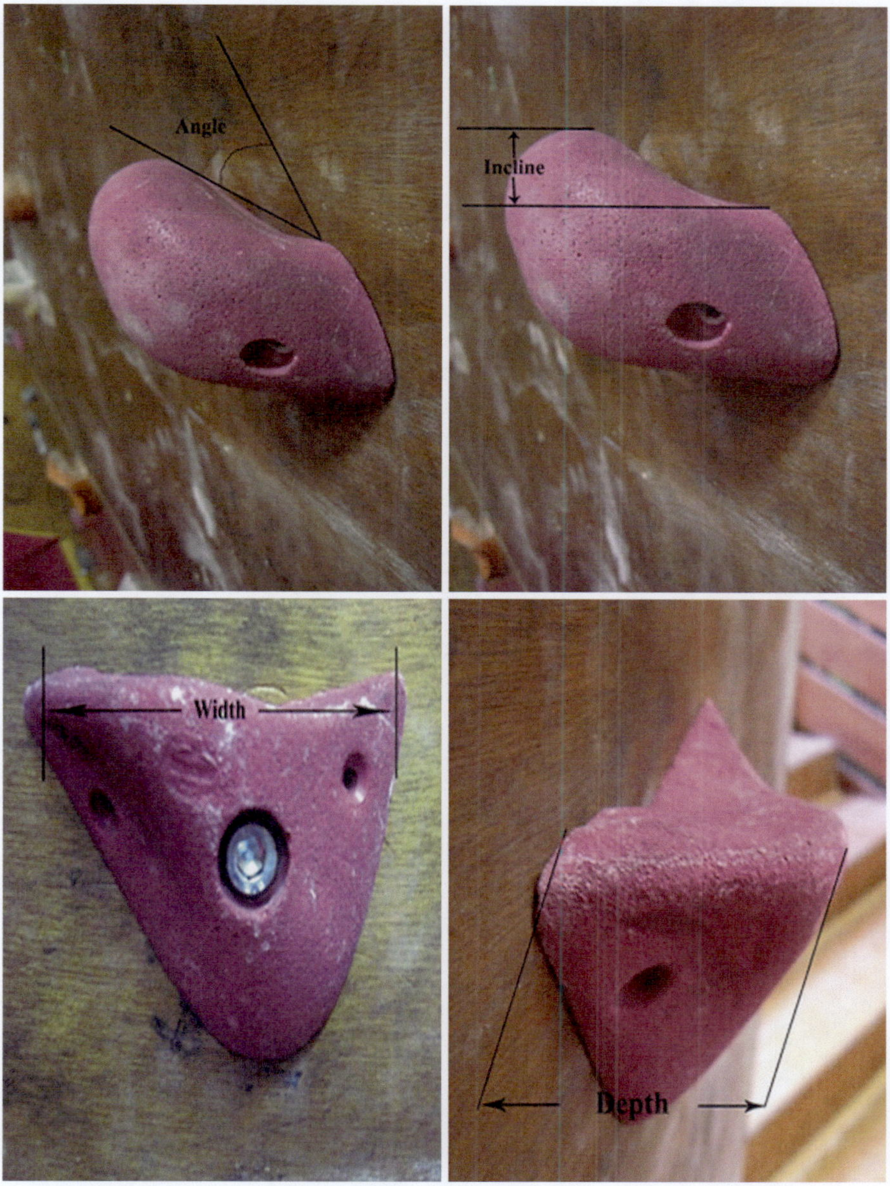

Photo 5.2 Fingers' action and position influencing factors

5.1.1.1 Open Hand Posture

This posture is not natural for a fresh climber. Indeed, the idea of "seizing the hold" refers to the image of "clinging to something" which is implied in the crimp grip.

Therefore, the open hand posture requires an adjusted training time, because the climber has to maintain a perpendicular pressure to the hold, which is not so obvious according to the available surface. The pressure results from the part of the finger or the hand which is in contact to the hold. The pressure felt by a straight finger is higher on the last phalanx that is to say on P3. In that posture, the finger flexes the DIP joint with a slight flexion of the PIP joint too. According to L. Vigouroux, thanks to the open hand posture, the strain endured by the pulley is reduced – 32 times lower on A2 and 3 times on A4.

Quoting Grant and Al., Sweizer, Quaine, and Vigouroux, "considering an identical hold, no strength differences between the two postures have been reported whether at the highest intensity – used on the hold – or at the lowest point when finger muscles get tired." Therefore, according to the shape, the size, the angle, the incline degree, and of course the climber's skills, it is advised to use the open hand posture as much as possible.

Photo 5.3 Strength during the slope grip (Illustration inspired by Duval, *La Main du grimpeur: approche physiologique, clinique et expérimentale*, 1986)

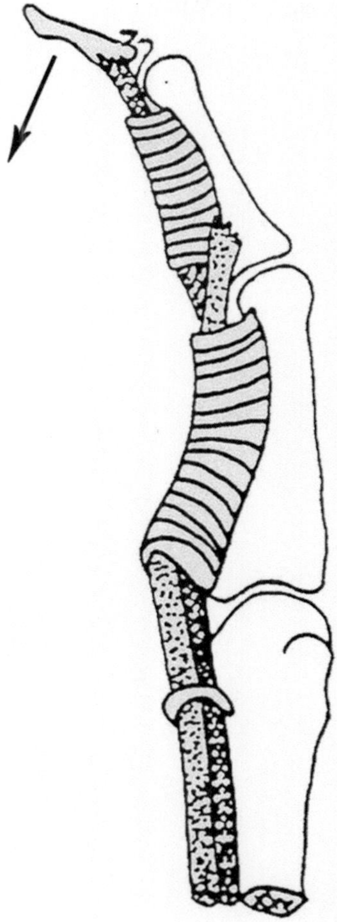

Schema 5.1 Strength during the slope grip (Illustration inspired by Duval, *La Main du grimpeur: approche physiologique, clinique et expérimentale*, 1986)

The strength over the two tendons FDS and FDP is equally distributed in the open hand posture. EDC strain is another parameter to be considered. For now, a specific training for these two muscles has not been reported yet. Coaches and climbers have to pay attention to the involvement of the EDC and see how to reinforce them.

Most of the time, the open hand posture requires the use of the four long fingers for large holds, which allows better contact with the surface in grips. As training goes by, the climber is able to use this posture on locked or rounded holds, adjusting the necessary number of fingers according to the size of the hold or the available surface. The open hand posture enables the climber to broaden his skills, using the four long fingers or just one finger to seize the holds.

Concerning the two-finger posture, it is more advised to use the middle finger and the ring finger. Indeed, the open hand posture maintains a better structure stability and limits the possible lesions on the pulleys in crimp grip. The middle finger prevails, and it is supported by the ring finger. These two fingers are supposed to be the strongest (Duval 1986), but they are also the most currently injured (Quaine et al.). The location of these two fingers in the longitudinal axis of the hand makes them the most convenient to be used.

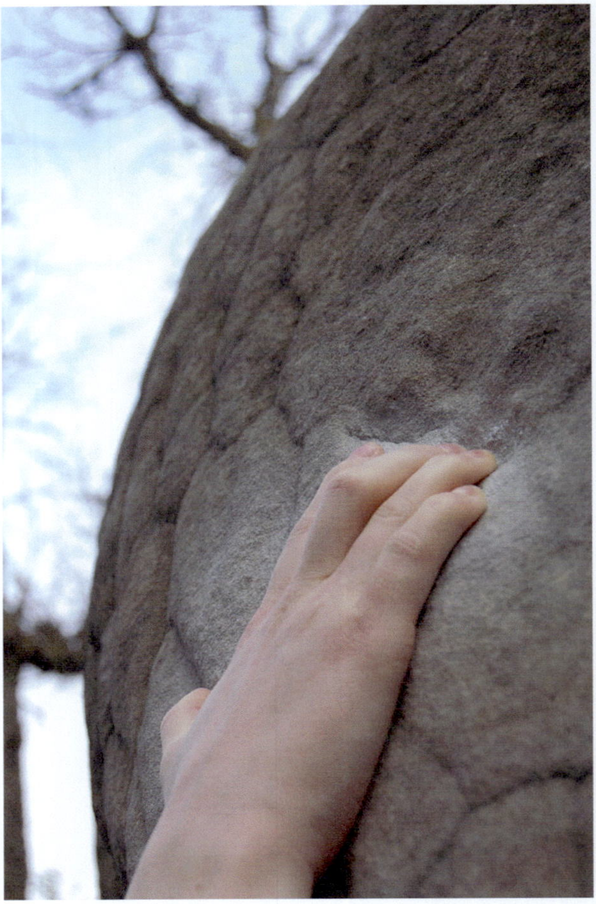

Photo 5.4 The slope grip

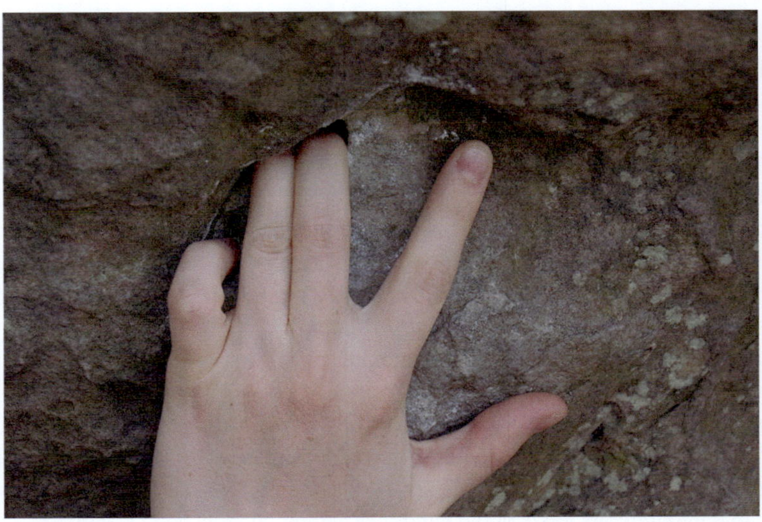

Photo 5.5 Two fingers in slope grip position

Tendinitis and tenosynovitis are the most current injuries resulting from the open hand posture. Most of the time, tiredness or exaggerated training accounts for these lesions. Some grips on steep holds or non-ergonomic supports tend to cause this kind of trauma.

Photo 5.6 Slope grip in a broad surface

5.1.1.2 Crimp Grip

This posture which is associated to clinging seems to be more natural for the beginner climber. For the advanced climber, the choice of this posture will depend on the width of the surface of the hold. The smaller and narrower the surface is, the more appropriate crimp grip is – with one or several fingers – even if the hold is flat.

In addition to that, this posture is more convenient than the open hand posture to reach a farther hold – due to the wrist's position which is higher.

Photo 5.7 Strength during the crimp grip (Illustration inspired by Duval, *La Main du grimpeur: approche physiologique, clinique et expérimentale*, 1986)

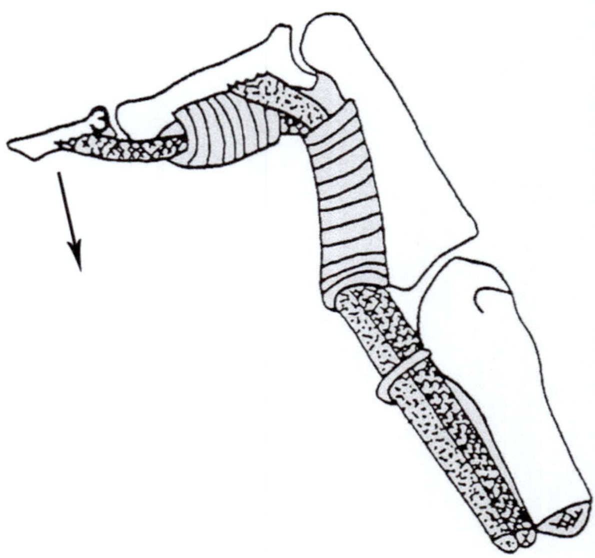

Schema 5.2 Strength during the crimp grip (Illustration inspired by Duval, *La Main du grimpeur: approche physiologique, clinique et expérimentale*, 1986)

In crimp grip, two cases are possible:

- The thumb is set on the index P3 phalanx. The pressure then brought on the phalanx is very strong on the surface level. But the pulleys are highly strained due to the angle of the PIP joint. In that particular case, the thumb acts like a finger locker to increase the strength.

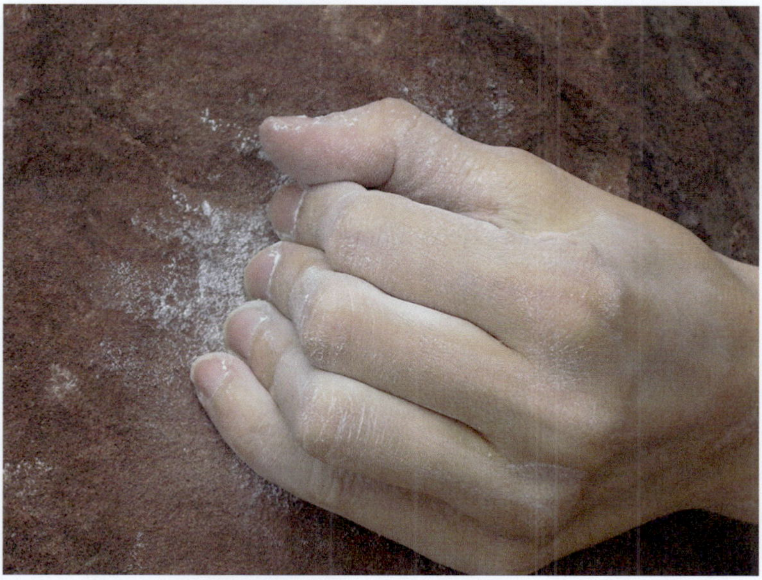

Photo 5.8 Crimp posture

- The thumb clings to the hold and opposes its strength to that of the fingers. This posture can be compared to that of the pliers grip though in the former case the strength involved is opposite. This posture is an intermediary between the open hand posture and the crimp grip posture, because the PIP joint angle is less broad and consequently the strength involved on the pulleys is reduced. This posture of the thumb is commonly used in CG, because the added holds on the wall make it possible, either on the edge of the hold or on a steep part, or a specific part of it, or simply on the screw hole – in competitions, some climbing leaders make sure to refill it, whereas some retailers reduce the screw hole so that only the head of the screw hole can be visible. In natural sites, this posture is not commonly used and depends on the bumps on the surface.

Photo 5.9 Crimp grip of the long fingers and pinch with the thumb

The crimp grip consists of a flexion of the PIP joint and a hyperextension of the DIP. Concerning the hardly hooked gripping – large holds – the pressure involved on the hold is focused on the P3 phalanx, which reduces the surface in contact.

This posture is a real strain on the A2 and A4 pulleys. According to L. Vigouroux, the A4 pulley is closer to its limit (85 %) than the A2 pulley (45 %). Indeed, the strength present on the A4 pulley is about 178.4 N, when its rupture point would be 210 N according to Lin et al. (1990). As regards the A2 pulley, the weight involved is about 200.2 N and its rupture point would be 465 N. The crimp grip posture is a real strain on the FDS muscle. FDP tendons and EDC extensors are active but to a lesser extent. Nonetheless, EDC activation is a factor to be considered as regards performances but also for the climber's health. The EDC muscle plays an important part.

The breaking off a pulley is the most common injury. In addition to the size of the hold, other factors increasing the tension of this fiber structure – angle of the hold, number of supports, etc. – or a too tiring training session or a bad preparation accounts for this kind of lesion.

5.1.1.3 Mixed Hand Posture

This posture involves a vast number of possibilities mixing crimp grips and open hand postures. For a great part and for each finger in action, this posture requires the flexor muscle DCF mobilization (in the crimp grip posture) or to an identical degree FDS, FDP, and the extensors EDC (in the open hand posture).

The mixed posture requires a more advanced level in climbing due to the specific strains endured by muscles. It can be quite traumatizing.

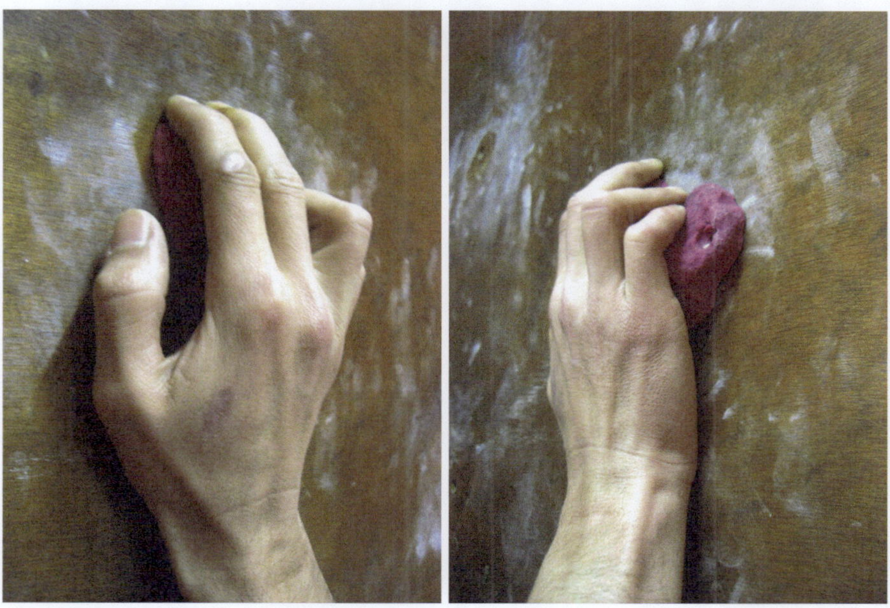

Photos 5.10 and 5.11 Combined grip position

5.1.1.4 Hooked Posture

This posture consists of a flexion of the PIP and DIP joints. It is a global hold, and it is usually used on large holds.

5.1.1.5 Locked Grip

This grip usually requires the open hand posture. The climber inserts his fingers inside a hole or a crack and rotates his forearm inwardly. This phenomenon of locking reinforces the strength in presence. This grip can be painful or traumatizing for the joints (PIP sprains). The open hand grip generally improves the locking movement in seizing and consequently involves on a same level SCF and DCF without forgetting the extensors FCE.

Photo 5.12 Locked grip

5.1.1.6 Fist Grip

This way of gripping is more frequent in CG than in natural sites. The interest of this posture is that for the great part the hand flexors are involved – the palm of the hand is in contact with the surface of the hold. In that precise case, the surface in contact with the hand is large enough, so that the finger common flexors are slightly relaxed. This grip is convenient for large and rounded holds. Then it is easier to place the hand on the side above the hold and clutch the fist.

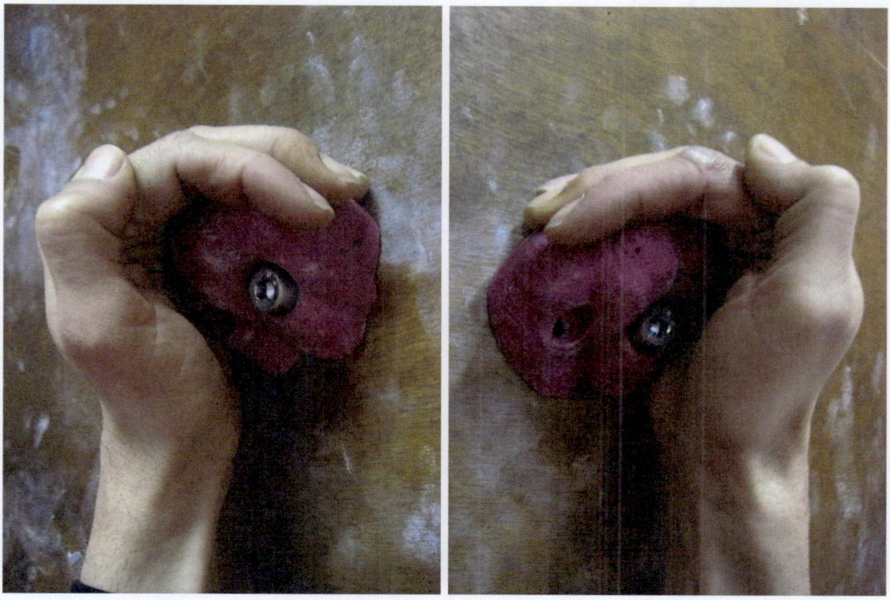

Photos 5.13 and 5.14 The knob grip

Photos 5.15 and 5.16 The knob grip

5.1.2 Crimp Grip or Open Hand Posture: Which One to Choose?

From beginner to advanced level, the climber improves his way of gripping discovering new and various holds. The fresh climber will limit his experience to the crimp grip posture, which is safer and more natural for him since the strength involved on the hold (hooked grip) allows him to be in contact to the structure.

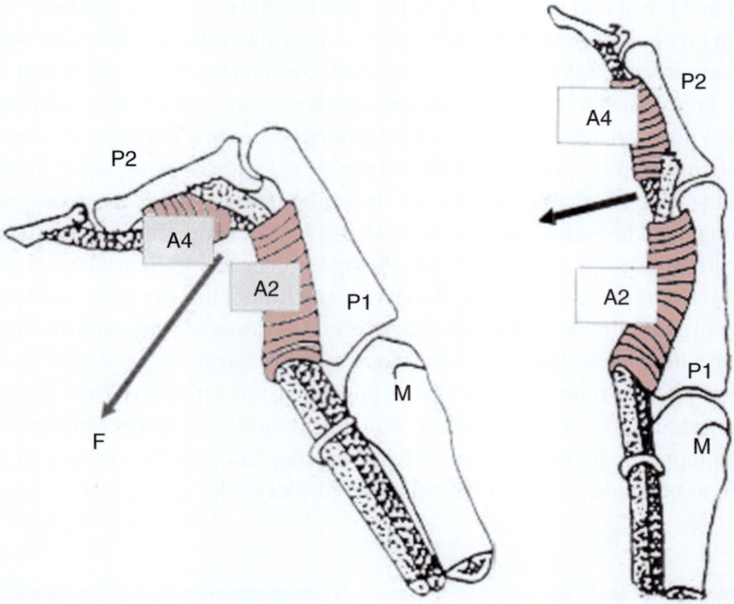

Schema 5.3 Different grip (crimp and slope) simplifications (Illustration inspired by Voulliaume (D.), Forli (A.), Parzy (O.), Moutet (F.), *Répartition des ruptures de poulie chez le grimpeur)*

As previously explained, the open hand posture is gradually achieved through a right adaptation and training to master the proper and different weight to be applied on the last phalanx of the fingers used (P3). In open hand posture, the other phalanges (P1 and P2) can also be in contact according to the size of the hold. This method stimulates at the same time the FDS and FDP, which requires a muscular reinforcement of FDS since the tendon is much more strained in open hand posture than in crimp grip – 165. N in open hand posture versus 113.2 in crimp grip. This kind of practice requires a gradual muscular fitting.

Does that mean that it is much better to climb in the open hand position? For sure the crimp grip is a stronger strain for pulleys – especially the A4 pulley on which the risks of rupture are higher – but it is also more convenient to seize the small and not too deep holds. Besides, according to the climber's hand morphology and level, the open hand position does not always help him to maintain a hold, which makes him quite uncomfortable. Indeed, a quite long middle finger or a too short little finger may be a source of trouble. Therefore, the enlightened climber has to adapt to his morphology and his skills, combining the open hand, crimp grip, or mixed position. In order to reduce the number of lesions, the climber has to be trained in all these different techniques. Nonetheless, the advanced climber has to put forward the open hand posture to limit the tension on the pulleys. Outdoors, during high-level competitions, and to achieve good results, fingers are excessively prompted. This is why open hand posture relieves some of the finger tension especially over the pulleys. After a finger injury, during a period of reeducation, this posture combined with a proper strapping is highly recommended. The adaptation to the crimp grip posture will have to be gradual, as will be explained further in this document.

Photo 5.17

For a climbing expert, the maximal intensity measured on a same hold is identical whether he uses the open hand or the crimp grip. It would be quite interesting to train the fresh climbers to use more often the open hand posture. Climber coaches have to bear in mind this phenomenon to preserve the beginners from finger injuries. The sooner a great variety of seizing holds is learned, the more aware and less injured the fresh climbers will be. The more trained the young climber is, the more experimental he is, and eventually he becomes more qualified to choose the righter and safer finger posture according to the situation throughout his future sports career.

Photo 5.18

5.2 Training Basics

This part of the book will deal with a few basic principles to avoid any mistakes during training practice. Further information concerning the different fields and the training planning can be found in many other books even if a bad training planning can lead to a lesion. A good planning aims at helping the sportsman or the competitor to be fit on a D day or at maintaining the climber to be outstanding during a certain period of time.

5.2.1 Intensity Principles

In order to avoid an exhausting training, the athlete has to adjust the quantity to the intensity of the work. During the physical preparation (PP), the training sessions have to be numerous so that the body may get used to effort and sustain the following sessions which will be more numerous and intensive. The more voluminous is the training session, the lower the intensity has to be, and reciprocally. It is dangerous to have enormous sessions in PP with routes close to the maximal level of the climber. During a specific preparation (SP) for a competition, the athlete's training sessions are focused on the specificities of the competition. The intensity is close to the maximum level or supramaximum, and the time to recover is longer. Indeed the body needs a certain amount of energy to try the route again.

The intensity principle has to be considered throughout the session. Warming-up is a priority to prepare the different muscles and joints, though a great number of climbers tend to forget it and prefer climbing gradually. Defining a specific time to get an optimal warm-up is quite difficult. It depends on the daily physical condition, the level of tiredness, and the aim of the session. Being aware of one's own body and paying attention to any physical signs are consequently extremely important.

5.2.2 Alternative Principles

Alternative principles can be divided into two categories:

5.2.2.1 Grip and Support Alternatives

The great number of possible positions makes climbing a varied and exciting sport, especially outdoor climbing which involves a great variety of grips and movements. Each climber has his own style with his strong and weak points. The climber has to train on various supports to be as efficient on any of them but also to avoid injuries. Working an unknown posture intensively may lead to a trauma. Climbing in different kinds of routes is a good way to broaden one's skills and to use the various joint and muscular chains and consequently to become an expert. As regards fingers, the climber has to be able to seize holds differently, especially if he wants to be highly competitive, outdoors or indoors. The repetition of the same movement is a high strain on the tendons, joints, and pulleys. Repeating over and over the same gesture remains a possible cause of injury.

5.2.2.2 Training Session Alternatives

Alternating route training, intensity and volume, is a good means to avoid psychological pressure. Indeed, if the athlete feels competent and determined and enjoys what he does, he will achieve more good results and the training sessions will be more efficient. The training sessions need to be varied including different drills even if the aim is identical. Emotionally speaking, it is generally hard for people to reproduce the same work for 3 days on. The lack of motivation may cause an injury. Giving a sense to each activity is a motivation factor. This is why the sportsman has to understand what he does to put a lot into the training session.

Alternative is the key word for a healthy body. The former helps the latter to remain in activity and to recover after a specific training session. Obviously, alternative is good for the body, but it is also a complementary way to deal with the emotional load resulting from the undergone efforts. Changing the training routes, telling himself that a hard session won't be repeated, and alternating the intensity or the volume of a session are the many means for the climber to take a deep breath.

To make the recovery easier, training cycles alternate between intensive and less intensive sessions. The issue in alternative would be to do the wrong thing. A logic principle has to be respected and depends on the different fields and the required time to recover. The training planning has to consider the notion of pleasure too.

5.2.3 Progressive Principles

The intensity and the volume of the training have to be reversed in order to avoid a habit in the body and to reduce possible trauma. After a full stop practice for whatever reasons – lesions, family matters, or personal reasons – it is dangerous to start practicing again at the highest level, because the body has more or less got out the habit of intensive efforts. Consequently, it is advised to start slowly and to increase gradually the intensity throughout the training sessions.

After an injury, the time to regain one's higher level is usually more or less double that of the rest period due to that lesion.

During a training planning and a regular practice, the principle of overweight is essential. To improve a prior factor of the performance such as the strength of the fingers, it is necessary to work on them by overloading them. For instance, the size of the grips, the number of supports, and the footholds can be reduced. Indeed, if the climber works only on good grips and comfortable footholds, he won't develop his strength. On the other hand, once back to training, if he overworks on a hangboard or on a campus board, he might get finger lesions.

Progressive principles imply the fact that training sessions have to be global and then specific. At the beginning of each season, the coaching of a lead climber and that of a bouldering climber of the same level tends to be similar. As the season unfolds, the sessions differ completely for each climber and the training becomes more specific.

The aim is to evolve and turn quantity into specificity. The climber reduces the number of route attempts and focuses on particular movements according to the objective of the session or the cycle.

If a climber or a coach wants to make a progressive training, before defining a training session, they have to think about the tiredness factor and its impact.

5.2.4 Constancy and Continuity Principles

Usually, interrupting climbing training whatever the reason and for more than a week is bad for a good preparation. Achieving good results, whatever the level, requires a more and more complete and demanding involvement. Going back to training is always a difficult time for climbers who do not perfectly know their bodies and their limits. For the most skilled ones, it is usually a frustrating period, because they are so eager to be at the top again that they often tend to do away with reeducation phases. An interruption – even if it has been only for 2 weeks – implies a gradual return to practice in order to avoid any injuries or recurrences. Several factors such as non-strained fingers or a not complete healing enhance the possibility of a new wound. A gradual return to practice is consequently highly recommended especially in case of former lesions. Initially, fingers have not been made for rock climbing; this is why they have to be protected.

A practice interruption involving no injury is to be banned, especially for someone preparing for a competition or a precise aim. This kind of interruption requires a change in the training planning to find the right balance between the daily activities to be fit for the D day.

After an intensive cycle and to make the recovery easier, it is healthier for the body to undergo a couple of less intensive sessions and to insist on the notion of pleasure or technique rather than stop practice completely. In case of injuries, the interruption must not be definitive. In fact, to maintain one's form, a muscular training has to be followed including working out, electrostimulation, jogging, etc.

Photo 5.19

5.2.5 Specificity and Characterization Principles

Climbing training has to be specific according to each individual. This principle has been widely reported due to the fact that each athlete responds differently to a given weight. Indeed, on the one hand, for an identical weight, the amount of work can be sufficient for an athlete, but on the other hand, it can be over the top for another athlete. Quoting Weineck, "Such a training method is perfect for one but is an extra work for the other one."

The quality of training depends on characterization. All of the sportsmen do not have the same physical capacities, needs, experiences, levels, or objectives.

Several factors have to be considered:

- Age
- Psychomotor development level
- Sports level
- Gender
- Past experiences
- Objectives
- Special field

The coach has to take a closer look at the climber's social life. The latter's love life, work, or future exams are important facts to be taken into account for the training planning.

5.2.6 *Knowing and Listening to One's Body*

This principle is the most complex of all the others, because only the athlete can tell how he feels and notice any pain in his body. The coach is supposed to detect any changes in the athlete's state of mind – a feeling of desire, tiredness, or pain – but he can't measure the intensity of them. It is a priority to understand that pain is a signal sent by the body to protect itself. When there is a pain, there is an injury – more or less serious. The sportsman has to tell the difference of the "good pains to the bad ones," more precisely the ones felt in the forearms when the lactic acid is flowing.

Several parameters have to be considered:

- *Physical tiredness* due to intensive sessions, cycles, and/or extra activities outside training practice. As regards young people, the concern lays in PE or their social life which can bring an excess of tiredness.
- *Moral tiredness* results from the intensity of the last few sessions (or the last one), a lowering in motivation, or other external factors such as school/work tiredness, family matters, or social life problems.

This is why it is extremely important to talk to each athlete and help him to have a better understanding of his body. The coach has to adjust his training session according to the athlete's present state of mind. The sportsman's degree of concentration has to fit the required activities.

Photo 5.20

5.3 Climbing Various Efforts

Training with specific tools such as a campus board or a hangboard requires high skills to avoid any lesions. These appliances are a real strain for fingers which may endure extra weight if exercises are not well adjusted. The climber has to be precisely aware of what he is working on, because according to the field – alactic anaerobics, lactic – the effort times and the rest periods will change. If you do not rest properly after a training session based on strength, the previous effort will turn into an effort on resistance and therefore increase the risks of lesions.

In this book which deals with hand injuries, it is obviously necessary to mention a few things about the different kinds of fields so that possible lesions due to bad training practice can be reduced. Nonetheless, the subject being wide, it is hard to elaborate on the training topic, and further information will be found in other documentations.

5.3.1 Alactic Anaerobic Efforts

This effort consists of very intensive exercises on a maximal power. This category is entitled to work on strength liberating a great source of energy in a short time. During the session, adenosine triphosphate (ATP) and also a small amount of creatine phosphate (CP) are used. When they are solicited at their maximum level, the very short reserves are used out after 7 s. This lapse of time is consequently a reference time in rock climbing practice. Climbers often mention the number of movements fulfilled. It is much better to work on the notion of required effort, which is more understandable. According to the movement, the quantity may change.

The aim of the training is to remain focused on the available energy and use it efficiently.

Alactic Anaerobic Intensity Recap

Intensity: maximal or supramaximal (>100 %).

Duration of the effort: between 3 and 7 s.

Recovery: between 1 min 30 s and 3 min.

Repetition of the movement: about 10. The climber has to stop the repetitions when he notices a decrease in intensity.

Alactic Anaerobic Capacity Recap

Intensity: between 90 and 100 %.

Duration of the effort: between 7 and 15 s.

Recovery: between 3 and 8 min.

Repetition of the movements: from 6 to 10. If the climber notices a decrease in intensity, he can reduce the duration of the effort which allows a great amount of work.

5.3.2 *Lactic Anaerobic Efforts*

This category begins on the first second but on a very low intensity. The intensity is higher after 10 s. The strength in use is high but not maximal. The intensity grows from 30 to 45 s according to the required effort. The function of the training is to develop a resistance against "lactic poisoning."

Alactic Anaerobic Intensity Recap
Intensity: maximal.
Duration of the effort: between 15 and 45 s.
Break between two repetitions: from 30 s to 3 min in order to get a partial recovery of the basic potential.
Break between two series: from 5 to 30 min in order to produce another effort of the same level.
Repetition of the movements: between two and six. It is important to judge the intensity of the effort properly. The session has to be stopped when the climber is unable to make a new effort with a sufficient intensity.

Alactic Anaerobic Capacity Recap
Intensity: the intensity has to be 80 %.
Duration of the effort: from 45 s to 2 min.
Break between two efforts: 2 min.
Number of repetitions: between 5 and 8.
Recovery between two series: 6 min.
Number of repetitions: between 6 and 8. In order to work sufficiently, the climber can reduce the duration of the effort and increase the recovery time.

5.3.3 *Aerobic Efforts*

It is false to believe that aerobics is useless in rock climbing. Of course rock climbing practice does not really appeal to aerobics, but it is nonetheless quite appropriate especially as regards PP.

Lesions being the main topic of the book, aerobics is first of all a good means to prepare the body of the athlete for the next season. All the energetic processes are developed through aerobics. But above all, it is a means to prepare ligaments and tendons for efforts.

In climbing, an increase aerobic process improves the recovery between two efforts. It can be quite useful in bouldering competitions, especially when the climber is allowed 6 min of recovery between two route attempts. Aerobics is also a good means to evacuate lactic waste – due to muscle contractions – and to improve the cardiovascular system.

Photo 5.21

5.4 Various Competitions of Rock Climbing

Climbing competitions can be divided into three categories: speed, lead, and bouldering. In climbing terminology, a route or a boulder is done on sight, when no former attempt has been done except for the given time of observation. On the contrary, when a climbing leader performs successfully, it is called a "flash performance."

5.4.1 Lead Competitions

In climbing, lead category is the most well known. Competitions take place in CG. The aim is to reach the top of the route. Though a limited time is required, it is the reached height which determines the rank. Time is not used to choose the winner. It is just a way to prevent too long performances. In case of equally placed athletes after the first round, the jury takes into account the height reached in the former round. If the finalists are still placed first, a super final is organized.

5.4.2 Bouldering Competitions

In climbing, a boulder is a short passage. The aim is to achieve a great number of passages. Two kinds of competitions exist:

- A circuit which is set for France championships and international competitions. According to the competitions and the rounds – qualifications, semi-finals, or finals – climbers have to pass between four and six boulders within 4–6 min. Between two passages, climbers are entitled to a break corresponding to the time allowed for the passages. In those kinds of competitions, climbers are isolated even during the qualifications. They are ranked according to the number of passages they have succeeded. In case of equally placed athletes, the number of attempts is first counted, then the number of areas achieved successfully (with specific holds chosen by leaders especially after a first difficult part), and, finally, the number of attempts to reach the holds in these areas.
- A contest which consists of two rounds: qualifications and finals. During the qualifications, climbers are not isolated and have to achieve a maximum of the requested passages. Then the five best qualifiers take part in a flash or an on-sight final. In finals, climbers are allowed to make as many attempts as they want.

5.4.3 Speed Competitions

This practice is the oldest one. The aim is to reach the top of the route as fast as possible. These competitions are rare in France but commonly practiced in East countries.

Routes are climbed with a top rope. Competitions are divided into two parts: qualifications and finals.

5.5 Outdoor Climbing

Photo 5.22 (sur la droite)

Boulder and route difficulties are classified by grades, which correspond to the technical difficulty of the passage – height, slope, succession of holds, and sizes of holds. Marks remain biased and depend on each athlete's strong and weak points. The climbers' morphology can be an advantage or a disadvantage in some passages.

In France, route rating starts on 3 and stops on 9th degree. Each degree is divided into three parts: 6a, 6b, and 6c. There are also intermediary degrees – 6a+, 6b+, and 6c+. Several difficulty levels have been applied.

As regards bouldering categories, several ways of rating are available even in France.

5.6 Training

Climbing development has led to the appearance of new specific training supports, such as the hangboard and the campus board. Obviously, they are a good means to develop finger strength and/or finger resistance intensively, but due to the extreme finger strain they require, they can be the source of new lesions, more or less serious.

5.6.1 Safety and General Precautions

Numerous factors have to be taken into account before training with these specific tools. First of all, they are addressed to very high-level climbers. A beginner is not allowed to use one of these appliances because his body won't be able to endure the strain. These tools are actually designed for highly skilled climbers (6c/7a) who lack strength and would like to progress. But above all, they are addressed to athletes who would like to achieve excellent results in competitions, outdoors or on CG for fun.

Because of growth problems – especially as regards finger cartilages – *these sorts of tools can't be used by 16-year-olds,* that is to say, until the teenager is physically mature enough to undergo the possible side effects of intensive climbing practice. This stage is hard to recognize for sure, especially when each adolescent is different from the other, but at least the age limit is a good means to reduce risks. In the long run, finger-intensive training can lead to after effects which can turn a sportsman's career short. Adolescence is the right period to develop motor learning. But this stage can't be the same for everybody, since each teenager has his own specificity and consequently will develop according to his own phases or rhythm. At that age, training consists of learning a great number of climbing movements and facing various situations, focusing on the teenager's feelings in each case. In the second puberty phase, the muscular development and the higher capacity to understand and create motor designs are perfect to improve performances. This period is ideal to learn all the climbing-specific physical skills. Just like the use of the particular appliances, the specific skills require a precise adjustment. For instance, climbing with no foot is a good exercise before dealing with the hangboard or the campus board.

The use of this specific equipment is to be banned when one goes back to climbing practice, after a complete interruption or any lesion. The extreme strain requested by these tools can lead to a recurrence or a new injury for the climber. The body structure may not be well prepared, healed up, or strong enough to endure such a strain. These devices will be part of the specific training and won't be used before a proper reeducation and a few climbing sessions.

A gradual and complete warm-up is a good means to be more efficient during training sessions, but above all, it allows muscles and tendons to support the strain without getting injured. It has to be complete from generalities to specificities and has to last about 15–20 min. Besides, at the end of the training, it is important to stretch the muscles in order for them to regain their initial length. Repetitive contractions, especially concentric ones, tend to shorten the muscles which then become more fragile. It is a priority to bring them back to their initial size.

During a session, it is advised to combine climbing movements on artificial structures and the use of the specific equipment. This combination tends to enhance strength, without forgetting technical gestures, and mobilize the upper and lower part of the body. Consequently, it is important to make this association to get good results.

Photo 5.23 Crimp grip on a reverse hold

5.6.2 Training Safety

Training on this specific equipment requires a particular preparation due to the intense strain endured by the muscles and joints. This is why these tools can't be used at the beginning of a session, because they are a real strain not only on the flexors. The body needs a period of adjustment to deal with this kind of pressure. These specific appliances can be added during the second phase of the PP and will complete the specific training according to the category – bouldering or lead – during the SP. During the PP, the training is first global even if the types of exercises (training circuit) are supposed to reinforce muscles without neglecting the antagonist muscles to maintain a proper balance.

During the second phase of the PP, the use of these specific tools can be replaced by "no foot" exercises on a campus board, which prepare the muscles gradually to concentric and eccentric contractions. The athlete or the coach can introduce these tools little by little and for a short while, in the training.

In SP and according to the chosen objective, the preparation can be focused on these appliances by the coach. The exercises are so demanding that most of the time the athlete is close to his limits. The training sessions must not exceed 45 min or 1 h (warm-up included), and long periods of recovery between each activity or repetition have to be taken.

Whether in PP or SP, the cycle on these tools must not exceed 3 weeks and has to be followed by a recovery cycle in order to gain strength and to avoid any tendon traumas (flexors and extensors). Three sessions a week is the maximum number during a cycle. Between the sessions, the period of recovery changes according to the aim: to gain strength, 24 h; to develop resistance, 48 h; and to maintain strength, 72 h.

To reduce any possible injury, it is highly recommended not to carry extra weight, especially during strength cycles. In theory, working with additional weight might increase strength, but it is also a supplementary strain on the tendons which are first designed to support the body weight. Fingers are not designed to bear climbing weight. Consequently it is much better to work on various grips, focusing on a specific category and according to the objectives of each cycle or session. On the other hand, working on grips is a good means to work on one's strong and weak points. Using in turn the different holds (buckets, small edges, large holds, or round edges) and changing the fingers' posture (open hand or crimp grip) lead to a more global work on strength. To relieve tendon strain, it is advised to use the slope posture rather than the crimp grip.

5.6.3 *Training on Specific Tools*

5.6.3.1 Hangboard Sessions

Introduction

The hangboard is used to develop muscular abilities which are specific to rock climbing and precisely concern the different holds in suspension that can be reproduced with this equipment. This tool enables the climber to develop specific muscles and more precisely "no foot" training. Hanging work prevails and is a good help to emphasize a specific work on the duration of the holds. Moving on the hangboard is possible though quite limited. This training leads the climber to be more precise in his movements, since it helps the coordination and synchronization of the fingers' opening and closing.

The hangboard is not only a training tool, but it is also a device used to end the warm-up in a more specific way. In that particular case, the climber is fitter to start the ongoing activity, but he has to follow a gradual warm-up. Likewise, the climber uses first the comfortable holds (buckets) remaining shortly suspended, and then he gradually turns to smaller holds. It is strictly forbidden to hang on tricky holds or to make extreme movements, especially when he has not warmed up properly.

Possible injuries resulting from the hangboard not only affect the fingers but also the shoulders, elbows, or other parts of the body. The warm-up has to well prepare all the joints and the different structures of the scapular belt and the arms. The hangboard is an excellent tool, but if "the golden rules" are not respected, it can be extremely dangerous for the climber's health.

Here are a few rules to be respected:

- Never work loaded. The climber chooses smaller holds to work properly in the chosen category.
- Respect recovery time.
- Never use the crimp grip. This posture is a real strain for the fingers and is reinforced when the feet are not used.
- Train only if you are perfectly fit.
- Warm up all the joints, including the finger joints, before practicing.

Several types of exercises are possible with the hangboard, but we will focus on those improving the duration of the holds (dynamic efforts will be dealt with the campus board) and those concerning isometric efforts.

A training hangboard is chosen according to the holds which must not be "traumatizing" ("overusing"). A hangboard has to be deprived of too steep angles, and the posture of the fingers has to be convenient.

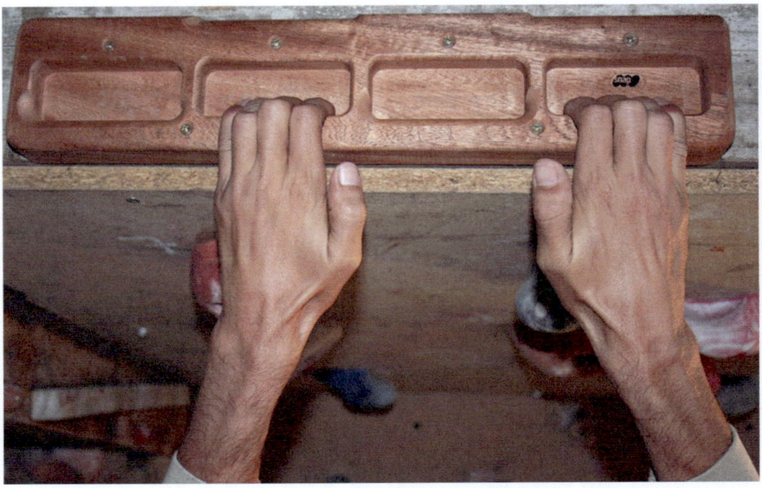

Photo 5.24 Training session with a wooden hangboard

Strength Development

After these kinds of sessions, the climber feels as if he has not done anything.

It is important to change the grips, in order to broaden one's skills and not to be limited to one particular way of seizing.

Principle: the climber hangs on holds, with one or two hands in the open hand posture. He is supposed to remain like that for 7 s maximum. To assess properly the intensity and if the hold becomes too easy, it is advised to reduce the size of it, even if the weight is reduced, rather than working loaded.

Duration of the effort: from 5 to 7 s.

Break between two efforts: 3 min minimum.

Repetitions: from 5 to 10.

Resistance Development

Training Session First Example

Principle: the exercise has to be fulfilled on the chosen holds: consequently, they must not be too small. In order for resistance to work, the climber has to remain suspended for at least 20 s.

It is possible to work with different degrees of closing arms.

Duration of the effort: between 20 and 45 s.

Break: 30 s.

Repetitions: between 5 and 8.

Break between two series: 5 min.

Number of series: 4.

Training Session Second Example

Principle: choose a hold to be able to remain suspended with one or two hands for 1 min.

Duration of the effort: 1 min.

Break between two repetitions: 2 min.

Repetitions: 3.

Break between two series: 5 min.

Number of series: 3.

5.6.3.2 Campus Board Sessions

Introduction

Just like the hangboard, the campus board is a powerful tool to work on the different anaerobics categories. Various movements, which can be fulfilled with one arm at a time or with both of them, develop dynamism, contact strength, and alactic anaerobic field.

Photo 5.25 Training session on a campus board

Safety regulations and recommendations are identical to those of the hangboard. This equipment requires a specific preparation such as the "no foot" on campus board, with a variation of holds and/or similar exercises. The climber can reproduce the same kind of effort using traction bars.

This appliance is ideal to work pliometric efforts. In that particular case, it is highly recommended to be cautious with finger structures which are much requested. Indeed, when the climber lets himself go on the lower hold, an extreme contact strength is involved. Once again, a specific preparation is compulsory.

Due to the extreme strain involved, it is advised to limit the number of sessions with a campus board. To begin with, one session a week seems to be sensible. Then as the climber becomes more skilled, he can practice two or three sessions a week maximum. The cycle must not exceed 3 weeks, so that the body can recover properly.

Strength Development

Training Session First Example: Dynamic Movements
Principle: choose holds on which the climber will be able to make up to five movements. The aim is to make movements farther and farther on smaller and smaller holds.
Number of movements: from 1 to 5.
Break between two repetitions: 5 min.
Repetitions: from 5 to 10.

Training Session Second Example: Dynos
Principle: choose holds on which the climber will be able to make a dyno with two hands. The aim is to reach a distant hold and when possible to make this movement in chain. The holds have to be large enough to reduce possible injuries.
Number of movements: from 1 to 5.
Break between two repetitions: 5 min.
Repetitions: from 5 to 10.

Resistance Development

Principle: choose holds on which the climber will be able to make at least ten movements.
Number of movements: the climber has to go up and down a campus board for at least ten movements.
Break between two series: 5 min.
Number of series: 5.

Photo 5.26

5.7 Return to Training After an Injury

Several reasons such as holidays, working life, or injuries can be the origin of an interruption in sports practice. In case of a lesion, it is quite different, and climbers and coaches have to be extremely cautious. Indeed, after an interruption, the body is not anymore used to the former practice. Therefore, it will react differently; physical abilities won't be the same, and consequently, the body won't be able to endure the former intensity and complexity required by the training. In that part of the book, we will deal with the basics of a return to training, bearing in mind that each session has to be characterized. The sportsman characteristics have to be considered, because each athlete is unique and so are his psychological state (motivations, objectives), physical abilities (skills, tiredness, recovery time, diet), anatomy (morphology, former injuries), and social life (leisure time, family, and working life).

Most of the time, the interruption is sudden due to an unexpected lesion; this is why activities have to be specific and varied once back to practice. During the interruption, family moral support is a priority to help the sportsman to accept his injury and recover faster. Physically, the dysfunction and sometimes the immobilization require a complete reeducation to the former automatisms. It is necessary to regain a good form, before focusing on a more specific preparation. The amount of work will be determined according to the athlete's form and abilities. Gradually, he will be able to perform at his rhythm to avoid any recurrences. Listening to the athlete, who of course will listen to his body, will be convenient to choose the right activities and to alternate properly intensive or important sessions.

During this readjusment period, what matters is not the energy involved of the dynamic movement performed but the way the body reacts (proprioceptive or kinesthetic information). The shocks resulting from the contact of the fingers on the holds, during dynamic movements or dynos, are the cause of injuries and consequently of recurrences. These contacts can create microlesions. Learning again all these movements and handling finger contact with holds have to be done gradually, in order to regain automatism in the synchronization of the fingers.

The climber feels completely recovered once he can grip holds strongly without feeling any pain. He needs to be convinced that the injury is healed up and that he can go back to practice safely. The readjustment period is necessary to improve the climber's confidence and to lead him to the next step of the return to practice. The climber needs to trust his body and his strength. This factor is important in order to use again the part of the body which was injured instead of trying to do without it. This reaction is quite common, for instance, the climber who hurts his left hand will unconsciously use the right one, which might eventually lead to new lesions.

The readjustment period is quite changeable and depends on various factors such as the seriousness of the lesion, the characterization of the training, and the sportsman's state of mind as regards the lesion. This is why the following part of the book will give general information as regards the return to training and the requested intensity of the sessions. Of course, it is not a universal solution, and each session has to be specific for each sportsman. The organization and the specificity of the sessions will be developed on general terms, since they will be different according to the climber's strong and weak points and according to the seriousness of the lesion.

This is why before the readjustment period, the climber and the coach have to analyze the causes of the lesion. This assessment is a good means not to make the same mistake again and to plan adequate sessions. For instance, in case of injury caused by a lack of hydration, the climber has to pay attention to that phenomenon and change his way of training. According to the cause or the kind of injury, the climber has to change his technique and his preparation – use the open hand posture, for instance, to relieve the pulleys.

5.7.1 Therapy

Readjustment period can be divided into different stages:

- *A stage including stability exercises.* They are made with strengths slightly higher than in daily life. For a finger lesion, the first stage consists of doing traverses keeping one's feet on the ground. The climber presses his hand gently, simulating the real movements he would have done if his feet had not been on the ground. He carries on the exercise, moving on the ground and adjusting the pressure on the different holds.
- *A stage including a daily use of the wounded part.* In case of a finger injury, the climber does traverses or passages, choosing carefully the holds so that he may not feel any unbearable pain. Big footholds relieve the body and reduce the finger pressure.
- *A stage aiming at returning to climbing practice gradually.* The climber makes passages quite inferior to his regular level so that he may get used to it and feel climbing sensations again.
- *A last stage aiming at regaining one's regular level.* The climber uses his finger with the requested strength to seize the holds without feeling any pain or any fear.

On-sight regular level and maximal level will be dealt with in that part of the book. A change has to be made according to the seriousness of the injury, which will lead to a more gradual return to practice as is the case after a pulley rupture. On-sight regular level will be associated to a mark obtained at least eight times out of ten.

The difficulty level will be measured in percentages. The maximal level (regular or not) equals 100 %. Then each letter is added a value of 10 %.

Let's take this as an example: the sportsman's on-sight regular level is 7a. Here are the corresponding values.

Photo 5.27

Level/rank	Percentages
7a	100
6c	90
6b	80
6a	70
5c	60
5b	50
5a	40

The following back-to-training planning is nothing more than simple examples. They have to be adjusted according to each sportsman's experience and can't be regarded as miraculous recipes!

Photo 5.28

5.7.2 *Tendon Injuries*

In that particular case, what matters is the seriousness of the injury which is determined by the duration of the interruption of practice. During this interruption, the athlete's level and capacities have obviously decreased, which implies a specific and different return to practice. Let's make the difference between a level 1 tendinopathy requiring less than 1 month of interruption and a level 2 tendinopathy requiring more than 1 month of interruption.

When the climber has been stopped for less than a month, the readjustment period is shorter. Thanks to a gradual return to practice, the athlete has to regain his initial level quite fast.

When the climber has been stopped for more than a month, after the readjustment period, he has to go through a PP to restart on good bases.

5.7.2.1 During the Interruption

During the interruption, the finger is not allowed to be flexed. Complete rest is compulsory. The athlete can nonetheless practice swimming, which is smoother as regards tensions.

During this period of interruption, the athlete can work his extensors and flexors by electrostimulation (reeducation program) and reinforcing the other muscles involved in climbing movements.

5.7.2.2 Level 1 Tendinopathy: Interruption Between 15 and 30 Days

First Week to Third Week

1. *Physical aspects*
   - The intensity of the effort must not exceed 60 and 70 % of the climber's on-sight maximal level (ML).
   - The training can't exceed three sessions a week.
   - On CG, the duration of the sessions is 01:30 maximum. They can last longer outdoors, but the intensity of the effort has to be decreased.
2. *Other aspects*
   - Alternating open hand posture and crimp grip posture is a priority to regain sensations when fingers are seizing holds. Be careful, the small holds have to be banned.
   - First of all, the training has to put forward moral and technical aspects and has to focus on climbing feelings and sensations, especially during the seizing of the holds. Throughout that time, the climber will have to trust his fingers to be sure of his complete recovery.
3. *Complementary training*
   - The athlete can complete his training with electrostimulation sessions (return to practice or reeducation program) three times a week.

Throughout this period, all the finger muscles have to be warmed up properly (using a rubber ball, modeling clay, or an elastic). Hydration (before, during, and after each session) and stretching exercises (finger deep flexors, extensors, etc.) are also extremely important. After each session, it is advised to ice the finger.

Fourth Week to Sixth Week

1. *Physical aspects*
    - The intensity of the effort must not exceed 70 and 100 % of the climber's on-sight maximal level (ML).
    - The training can't exceed four sessions a week.
    - On CG or outdoors, the duration of the sessions is 02:00 maximum.
2. *Other aspects*
    - The alternative and the variation of the different kinds of climbing and grips allow the climber to readjust his body to any situation.
    - Technical, tactical, and physical aspects have to be favored, with a muscular reinforcement of the other parts of the body.
3. *Complementary training*
    - Three times a week, the athlete can complete his training with electrostimulation sessions – strength or resistance program according to the athlete's field.

    Hydration, warm-up, and stretching exercises are the priorities during all the sessions.

    Return-to-training recap:

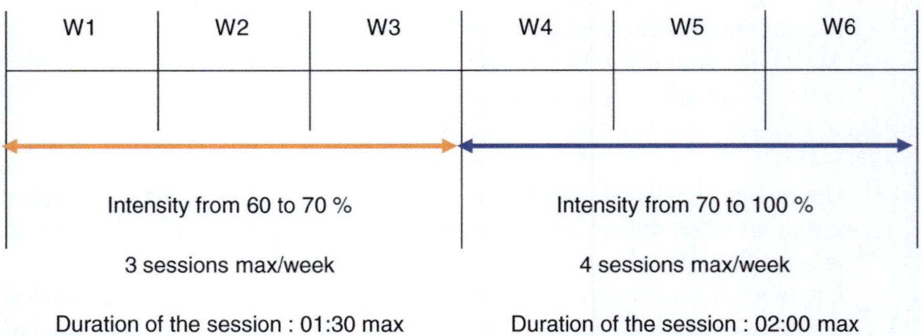

5.7.2.3 Level 2 Tendinopathy: Interruption Over 30 Days

First Week to Third Week

1. *Physical aspects*
    - The intensity of the effort must not exceed 50 and 60 % of the climber's on-sight maximal level (ML).
    - The training can't exceed three sessions a week.
    - The recovery time between two sessions is 48 h minimum.
    - On CG, the duration of the sessions is 01:00 maximum. They can last longer outdoors, but the intensity of the effort has to be decreased.

2. *Other aspects*

   - Alternating open hand posture and crimp grip is a priority to regain sensations when fingers are seizing holds. Be careful, the small holds have to be banned.
   - First of all, the training has to put forward the moral and technical aspects and has to focus on climbing feelings and sensations, especially during the gripping of the holds. Throughout that time, the climber will have to trust his fingers again to be sure of his complete recovery.
   - It is advised to avoid any dynamic movements or dynos.

3. *Complementary training*

   - Three times a week, the athlete can complete his training by electrostimulation sessions –return to practice or reeducation program.

The recommendations concerning the session are identical to those of a level 1 tendinitis.

Fourth Week to Sixth Week

1. *Physical aspects*

   - The intensity of the effort must not exceed 60 and 80 % of the climber's on-sight maximal level (ML).
   - The training can't exceed three sessions a week.
   - The recovery time between sessions is 24 h minimum.
   - On CG, the duration of the sessions is 01:30 maximum. They can be longer outdoors.

2. *Other aspects*

   - Alternating finger postures is still recommended, but the athlete can gradually reduce the size of the holds or the number of the footholds.
   - Little by little, the climber can learn dynamic movements again.

3. *Complementary training*

   - Three times a week, the athlete can complete his training with electrostimulation sessions – muscular strengthening program.

Seventh Week to Ninth Week

1. *Physical aspects*

   - The intensity of the effort must not exceed 80 and 90 % of the climber's on-sight maximal level (ML).
   - The training can't exceed four sessions a week.

- Throughout the week, the athlete can't train more than two times, one after the other.
- On CG, the duration of the sessions is 02:00 maximum. They can be longer outdoors.

2. *Other aspects*

- The alternative and the variation of the different kinds of climbing allow the climber to readjust his body to any situation.
- Muscular strengthening has to heighten technical, tactical, and physical aspects.

3. *Complementary training*

- Three times a week, the athlete can complete his training with electrostimulation sessions – strength or resistance program, according to the athlete's field or objective.

Tenth Week to Twelfth Week

1. *Physical aspects*

- The intensity of the effort must not exceed 90 and 100 % of the climber's on-sight maximal level (ML).
- The training can't exceed four sessions a week.
- Throughout the week, the athlete can't train more than two times, one after the other.
- On CG, the duration of the sessions is 02:00 maximum. They can last longer outdoors.

2. *Other aspects*

The recommendations concerning the session are identical to the cycle W7–W9. Return-to-training recap:

> *The athlete may need a couple of weeks to recover. These recovery weeks can be set between each cycle, that is to say, on the 4th week, 7th week, and/or 10th week. Of course, the next cycle will have to be postponed to the following week.*

Example:

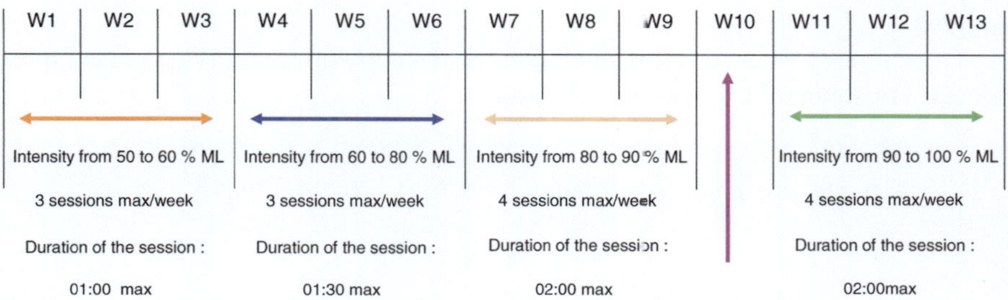

Photo 5.29

5.7.3 Digit Pulley Injuries

5.7.3.1 Digit Pulley Partial Rupture

After 30 Days of Interruption

The athlete can start a muscular reeducation by circuit training or muscle-developing exercises – without forcing the fingers to flex – completed with swimming and jogging.

Three times a week, electrostimulation (reeducation program) is a good complementary treatment for flexors and extensors.

> *Climbing and any finger work against resistance are prohibited.*

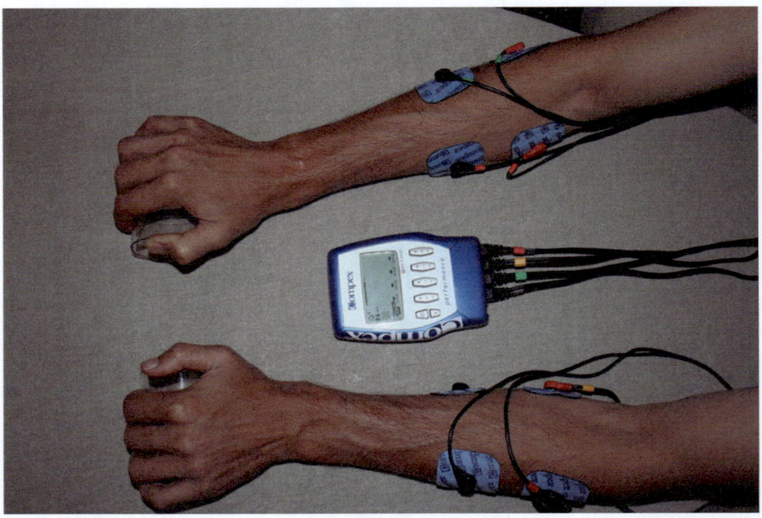

Photo 5.30 Flexor and extensor tendons electrostimulation session

After 45 Days of Interruption

After such a lesion, the climber has to become aware of the importance of changing his habits as regards hydration, warm-up, and stretching exercises.

First Week to Third Week

1. *Physical aspects*
   - The intensity of the effort must not exceed 40 and 50 % of the climber's on-sight regular level (RL).
   - The training can't exceed two sessions a week.
   - The duration of the sessions is 01:00 maximum.
   - The recovery time between the different passages have to be long.
2. *Other aspects*
   - Throughout that period, the best thing to do is to train outdoors as much as possible, choosing the adequate climbing style. Round holds are essential to help the gradual return to sensations. Alternating grips improve the relearning of the hand gesture.
   - The open hand posture has to prevail during the seizing of the holds. The crimp grip posture won't be used exhaustively by the third week.
   - The focus is on quality, insisting on techniques, tactics, and psychology.
3. *Complementary training*
   - Three times a week, electrostimulation sessions – back to training or reeducation program – can complete the training.
   - During the day, flexors and extensors can be strengthened with a rubber ball or an elastic. Using the former injured finger to squeeze the ball, in the crimp grip posture, is forbidden. Actually, the ball is set in the middle of the hand, and all the fingers rub it.

Throughout this period, all the finger muscles have to be warmed up properly – using a rubber ball, modeling clay, or an elastic. Hydration (before, during, and after each session) and stretching exercises (finger deep flexors, extensors, etc.) are also extremely important. After each session, it is advised to ice the finger.

The sore is always present during the first 2 weeks. The sessions have to be focused on quality. And the psychological side prevails to help the climber deal with the pain on his return to practice.

Fourth Week to Sixth Week

1. *Physical aspects*
   - The intensity of the effort must not exceed 50 and 70 % of the climber's on-sight regular level (RL).
   - The training can't exceed three sessions a week.
   - The duration of the sessions is 01:30 maximum.
   - The recovery time between the different passages have to be long.
2. *Other aspects*
   - Throughout that period and according to his capabilities, the climber can alternate CG or outdoor sessions.
   - The open hand posture has to prevail during the seizing of the holds. Nonetheless, during the session, the crimp grip posture can be used on specific tasks.
   - Crimp grip is practiced on large enough holds allowing the use of the slope grip posture.
   - The work is principally based on quality focusing on technique, tactics, and psychology.
   - From the sixth week, the athlete can practice volume training of low intensity (40 %) to practice a little more climbing. The open hand posture has to be emphasized and tiredness has to be taken into account.
3. *Complementary training*
   - Three times a week, electrostimulation sessions – muscle strengthening program – can complete the training.
   - During the day, flexors and extensors can be strengthened with a soft ball or an elastic.

Seventh Week to Ninth Week

1. *Physical aspects*
   - The intensity of the effort must not exceed 70 and 90 % of the climber's on-sight regular level (RL).
   - The training can't exceed three sessions a week.
   - The duration of the sessions is 02:00 maximum on CG. They can be longer outdoors.
   - The recovery time between the different passages have to be long.
2. *Other aspects*
   - Open hand posture and crimp grip have to be alternated. The aim of this cycle is to regain sensations in open hand posture, as much as possible. The athlete

may have to change his way of climbing to focus on this posture. Crimp grip is practiced on large enough holds, allowing the use of the slope posture.
- The work is principally based on quality rather than quantity. Muscular strengthening is another objective of the cycle.

3. *Complementary training*

   - Three times a week, electrostimulation sessions focused on the flexors and extensors –strength or resistance program – can complete the training, according to the climber's field or objective.

Tenth Week to Twelfth Week

1. Physical aspects

   - The intensity of the effort must not exceed 90 and 100 % of the climber's on-sight regular level (RL).
   - The training can't exceed four sessions a week.
   - The duration of the sessions is 02:00 maximum on CG. They can be longer outdoors.
   - The recovery time between the different passages have to be long.

2. *Other aspects*

   The recommendations concerning the session are identical to the cycle W7–W9. Return-to-training recap:

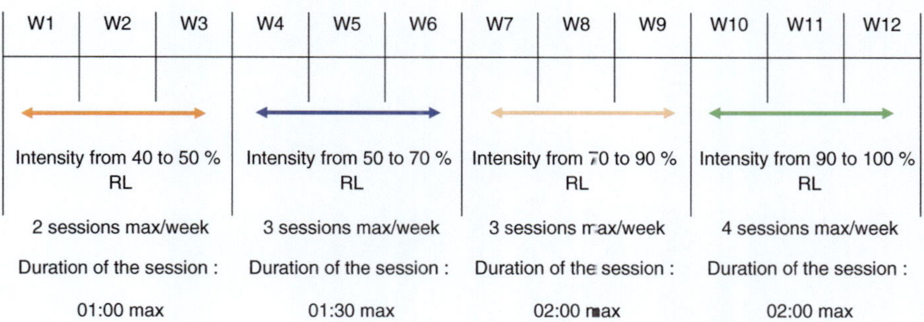

The athlete may need a couple of weeks to recover. These recovery weeks can be set between each cycle, that is to say, on the 4th week, 7th week, and/or 10th week. Of course, the next cycle will have to be postponed to the following week.

Example:

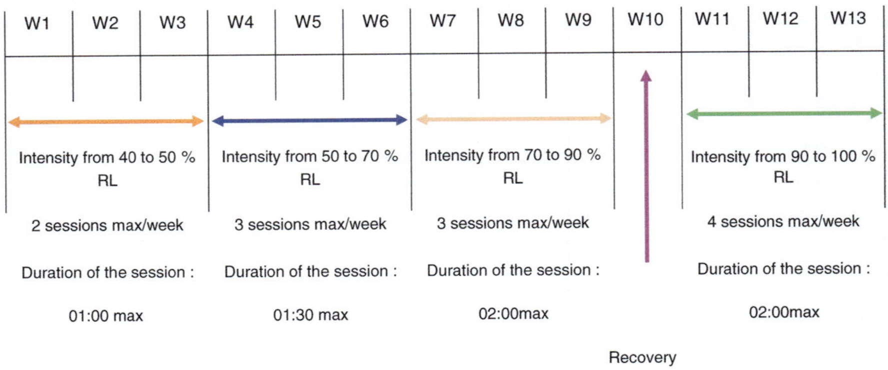

Photo 5.31

5.7.3.2 Return to Training After a Pulley Rupture Surgery

This kind of injury requires a great patience from the athlete. According to us, even if the recovery of the initial level is achieved in 80 % of the cases, the reeducation is long and painful especially at the beginning. For that sort of lesion and more particularly during the postsurgery period, the finger must not be flexed. Therefore, during the interruption (45 days after the surgery), the climber has to practice other physical activities. *But climbing and any finger work against resistance are forbidden.* Any physical activity requesting finger squeezing with devices such as a racket, a ball, or a barbell is not allowed. The finger must not undergo any shock or pressure to enable a gradual and perfect healing.

During that reeducation period and up to 90 days after the operation, the physical aspect of the training aiming at developing finger strength is banned. All the exercises requiring the use of the hangboard or campus board are strictly forbidden.

It is advised to return to practice *outdoors*, where the grips are more rounded and allow a balance of the weight on all of the fingers, including the palm of the hand. Long efforts (routes), rather than intensive ones (boulders), will be favored. *Reeducation in cliff routes seems to be the best way to readjust oneself to effort.*

Here is an example of a reeducation planning. The following information is not exhaustive and can be changed. Most of the time, duration and maximum intensity, not to be exceeded, are mentioned so that safety and healing are assured. All the information can be adjusted to a lower level, according to the athlete's physical form, state of mind, and of course pain felt during the exercises.

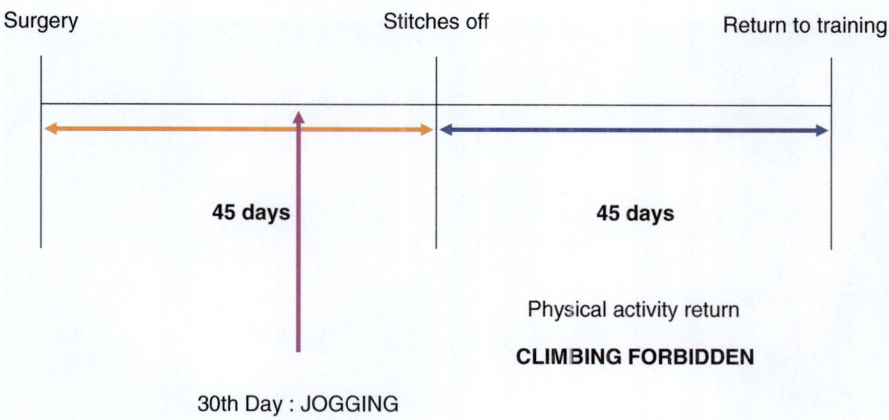

During the Interruption

A couple of exercises are suggested by D. Thomas. They can be done after surgery – from the 45th day to the 90th one – to reduce the after-surgery edema.
Exercise 1:

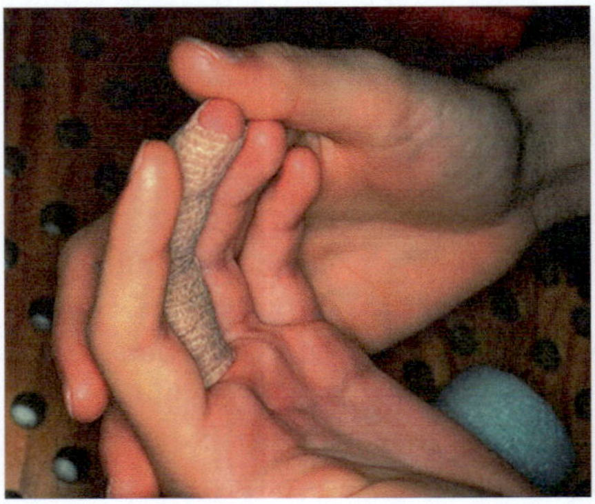
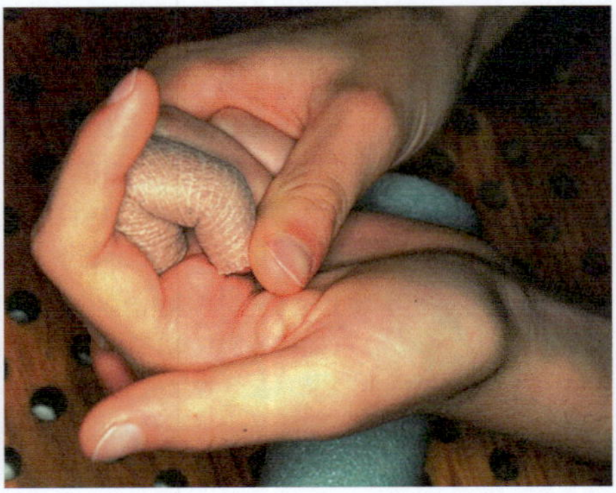

Photos 5.32 and 5.33 Total passive flexion of the repaired finger

Exercise 2:

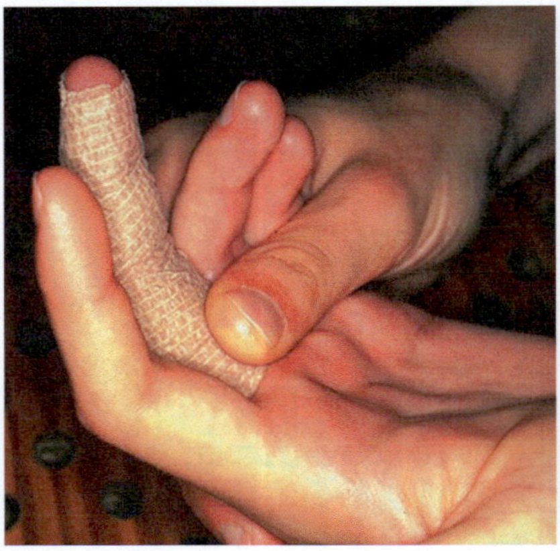
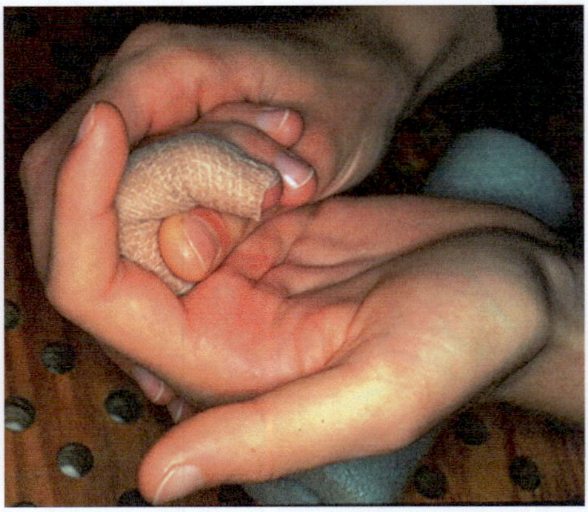

Photos 5.34 and 5.35 Protected active flexion. The physiotherapist's thumb supports the repaired pulley

Return to Practice: 90 Days After Surgery

The return to climbing has to be gradual and smooth to help the specific reeducation of that kind of injury.

First Week to Third Week

1. Physical aspects

   - The intensity of the effort must not exceed 20 and 40 % of the climber's on-sight regular level (RL).
   - The training can't exceed two sessions a week. The recovery time between two sessions is 48 h minimum.
   - The duration of the sessions is 30 min maximum, but it also depends on the pain previously felt during the first sessions. The athlete has to pay attention to it and must not overstress the former injured finger. The first sessions can be quite short and the climbing part quite limited.
   - The recovery time between two passages has to be extremely long (5–10 min).

2. Other aspects

   - The open hand posture is highly recommended. At the beginning, the crimp grip posture is forbidden to allow the athlete's body to readjust gradually.
   - Sensations and moral aspect have to be put forward by the coach and the athlete. Indeed, the athlete has to trust his hand again and believe in the success of his graft. This phase, which is not automatic, is necessary to allow the athlete to deal with the next step of the return to training. The climber and the coach have to pay attention to the right balance of the different strengths. After such a lesion, it is not unusual to see the athlete overuse the limb which hasn't been injured. This behavior may lead to another injury due to the extra weight undergone by the fingers. Special attention has to be given to the choice of routes and climbing styles, to enable the climber to use the former injured limb efficiently and safely.
   - It is not advised to practice on small edge slabs, such routes might be a real strain for fingers, just like overhangs by the way. This is why the choice of the right climbing style is a priority once back to practice. It is better to select vertical routes with large holds or buckets.

3. Complementary training

   - It is interesting to work on finger proprioception while swimming. Crawl and breast stroke are good exercises to strengthen the fingers in an environment not too traumatizing.
   - Three times a week, electrostimulation sessions – return to training program – can complete the training.

- A work on the extensors can complete the physiotherapy sessions to adjust the balance between all the fingers. This activity can be performed three times a week for 30 min.
- Throughout the week, the good thing to do is to rub a soft ball, which has to be as large as the climber's hand.

Below is a classic example of a training week planning:

Monday	Tuesday	Wednesday	Thursday	Friday	Saturday	Sunday
Climbing	Electro-stimulation	Physiotherapy	Break	Climbing	Electro-stimulation	Physiotherapy

Throughout this period, all the finger muscles have to be warmed up properly – using a rubber ball, modeling clay, or an elastic. Hydration (before, during, and after each session) and stretching exercises (finger deep flexors, extensors, etc.) are also extremely important. After each session, it is advised to ice the finger.

It takes a long time for the pain to disappear, and sometimes it is still there several months after the return to training. The way the climber feels on open hand posture is essential. Finger muscle strengthening is not the first priority for the climber.

Fourth Week to Sixth Week

1. Physical aspects
   - The intensity of the effort must not exceed 40 and 60 % of the climber's on-sight regular level (RL).
   - The training can't exceed three sessions a week.
   - The duration of the sessions is 01:00 maximum.
   - The recovery time between the different passages has to be long.
2. Other aspects
   - The open hand posture is still highly recommended.
   - Sensations and moral aspect have to be put forward in those sessions, since they are – just like the first period – the main points as regards the athlete's "reconstruction."
3. Complementary training
   - Three times a week, electrostimulation sessions – return to training program – can complete the training.
   - Besides physiotherapy, a work on the flexors and extensors is advised.
   - Swimming sessions are still pertinent.

Seventh Week to Ninth Week

1. Physical aspects

   - The intensity of the effort must not exceed 60 and 80 % of the climber's on-sight regular level (RL).
   - The training can't exceed three sessions a week.
   - The duration of the sessions is 01:30 maximum. This lapse of time can be adjusted according to the athlete's form and the pain he may feel.

2. Other aspects

   - The sessions are focused on the changing of grips emphasizing on the open hand posture. The athlete has to broaden his climbing gestures by practicing on different slopes or routes – large edges, small edges, but no two-finger or one-finger pocket postures.
   - The climber can train 2 days, one after the other, maximum. A recovery period of 24 h is necessary after two sessions.

3. Complementary training

   - Three times a week, electrostimulation sessions – muscle strengthening – can complete the training.

   Below is a classic example of a training week planning:

Monday	Tuesday	Wednesday	Thursday	Friday	Saturday	Sunday
Climbing	Electro-stimulation	Break	Climbing	Break	Climbing	Electro-stimulation

Tenth Week to Twelfth Week

1. Physical aspects

   - The intensity of the effort must not exceed 80 and 90 % of the climber's on-sight regular level (RL).
   - The training can't exceed four sessions a week. The athlete can't train more than 2 days one after the other.
   - The duration of the sessions is 02:00 maximum.

2. Other aspects

   - The sessions can be focused on tactics and techniques.
   - Volume sessions can be organized, but the intensity has to be slightly lowered to 60 %.

3. Complementary training

   - Three times a week, electrostimulation sessions – muscle strengthening – can complete the training.

Thirteenth Week to Fifteenth Week

1. Physical aspects
   - The intensity of the effort must not exceed 90 and 100 % of the climber's on-sight regular level (RL).
   - The training can't exceed four sessions a week.
   - The duration of the sessions is 02:00 maximum.
2. Other aspects
   - The recommendations concerning the sessions are identical to the cycle W10–W12.

Information/Recommendations

Throughout this period, the intensity of the work depends on the athlete's capacities and the way he has accepted his injury. If the climber has not recovered his former level, it is advised to emphasize a volume cycle, with a session in the week when the climber will be able to try more difficult passages.

Below is a classic example of a volume week planning:

Monday	Tuesday	Wednesday	Thursday	Friday	Saturday	Sunday
Volume	Volume	Break	Volume	Break	Volume	Break
60 % max	50 % max		60 % max		50 % max	

Return to training recap

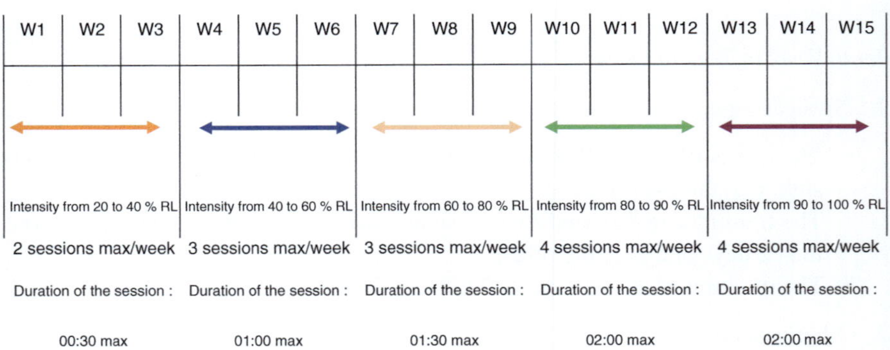

The athlete may need a couple of weeks to recover. These recovery weeks can be set between each cycle, that is to say, on the 4th week, 7th week, and/or 10th week. Of course, the next cycle will have to be postponed to the following week.

Example:

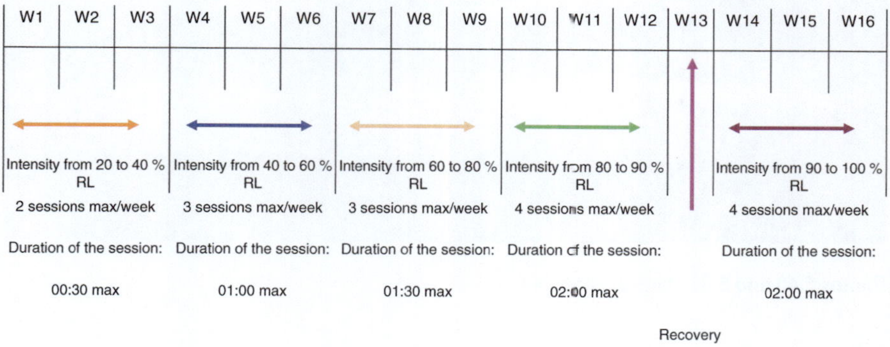

Strapping

The following example is a means to relieve A2 pulley.

The "tape" is 1 cm wide – as wide as a phalanx – and long enough to make a perfect bandage.
The "tape" is wrapped around the base of the first phalanx two times.
Under the PIP joint, the P2 phalanx is wrapped around two times.
The process is done a second time on the former wrapping under the PIP joint.

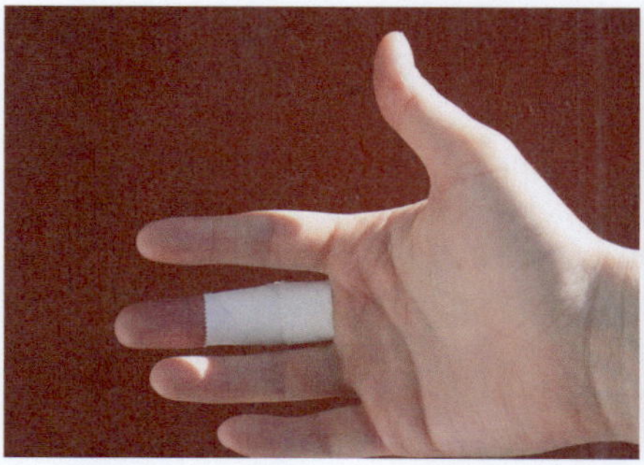
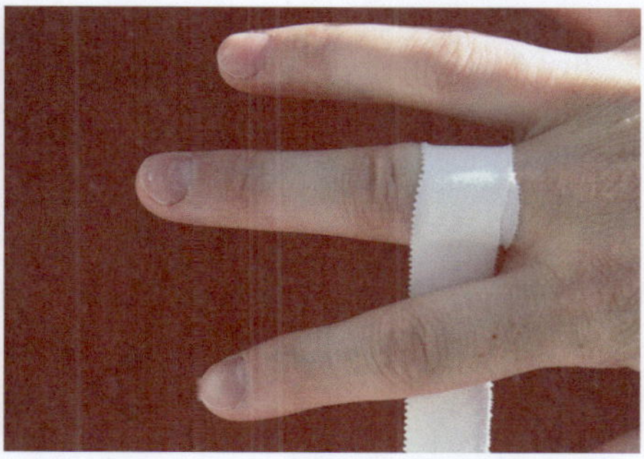

Photos 5.36 and 5.37

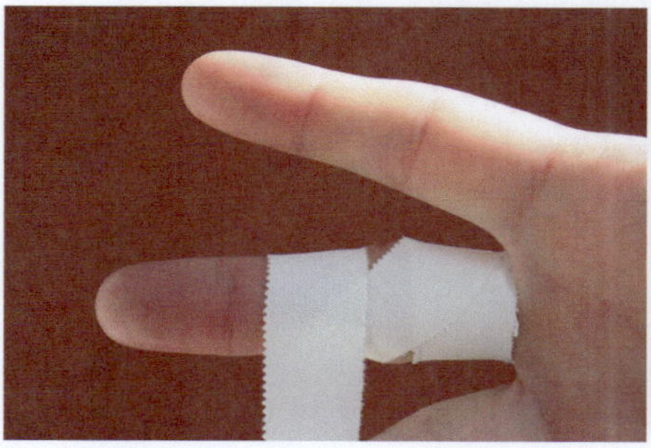
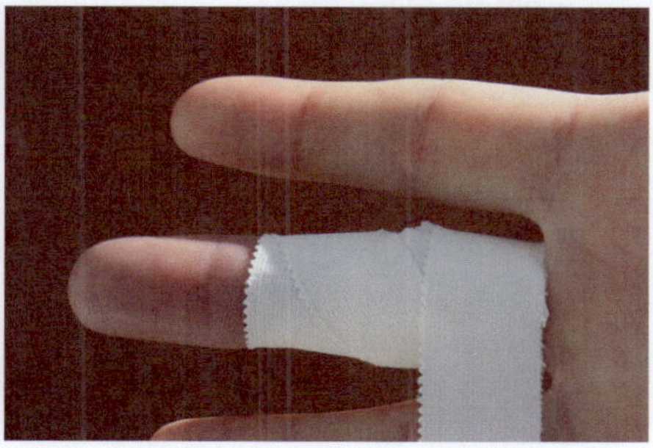

Photos 5.38 and 5.39

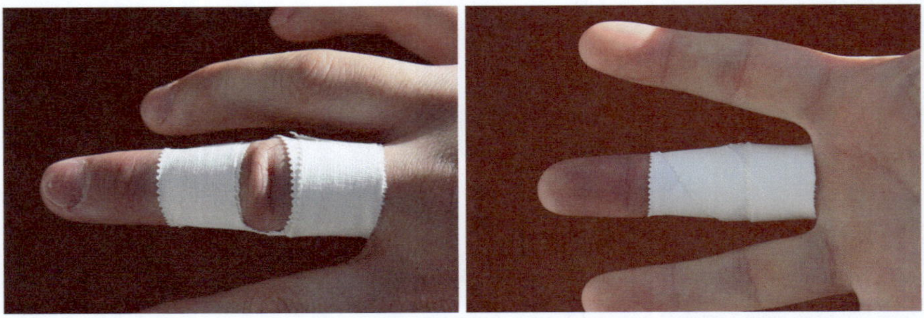

Photos 5.40 and 5.41 Digital strapping

5.7.4 Bones Injuries

The athlete can start a muscular reeducation by circuit training. A physiotherapist's advice and recommendations help a lot for a gradual recovery; this is why the choice of the exercises is a priority. Bone lesions usually lead to tendon stiffness due to the requested immobilization, which is quite an issue. Therefore, it is necessary to gradually prepare the fingers to climbing specifics. The physiotherapist's work aims at recovering a complete flexion and extension capacity.

Electrostimulation sessions (return-to-training program) based on flexors and extensors seem to be appropriate in that phase of reeducation to find the right balance between these two opposite movements.

Just like the other lesions, some grips are forbidden! The pressure undergone in two-finger or one-finger pocket postures is too intense, even if the climber does not use the former injured finger. Let's take the example of a ring finger fracture. If the climber uses his middle finger in a one-finger pocket posture, an important weight is automatically applied on the ring finger. During reeducation, it is important to keep one's fingers joint. Torsions on the injured fingers have to be looked after carefully.

Before going back on track, it is advised to be sure of the complete knitting of the injury. X-rays are the best means to confirm the diagnosis.

The causes of these kinds of injuries have to be found out, so that the climber may change his habits as regards the finger posture, the hydration, the warm-up, and the stretching exercises. After each session, stretching exercises are a priority to help promote recovery and enable fingers to regain their former position.

5.7.4.1 Level 1 Bone Injuries: Functional Treatment

First Week to Third Week

1. *Physical aspects*

   - The intensity of the effort must not exceed 40 and 50 % of the climber's on-sight maximal level (ML).

- Between two training sessions, a break of 48 h minimum is required. During the first week, it is advised not to do more than two sessions. As regards the second and third week, the climber can go up to three sessions maximum.
- The first sessions have to be short, 1 h maximum.

2. *Other aspects*

- Crimp grip is to be banned. To regain a good deal of sensations, different ways of seizing have to be used – large holds, buckets, small edges. Open hand posture is recommended to avoid any torsions.
- It is highly advised to avoid any tough contacts with the fingers and supports. Dynamic movements and dynos are to be avoided.
- Throughout the cycle, pains can be felt. After the session, icing the injured part of the body is soothing.
- Psychological aspect has to be emphasized. A great part of the sessions are focused on sensations and various situations.
- What matters is the way the athlete feels. Indeed, he has to trust his hand again and believe in the healing of his injury.

3. *Complementary training*

- The climber can complete his training with electrostimulation sessions if he does not make any with a physiotherapist. The program "return to practice" or "reeducation" allows a gradual return to practice.
- Throughout the week, the climber can use a rubber ball or modeling clay to regain mobility in finger flexion and extension.
- During that period, swimming is also a good means to reinforce all the structures gradually.
- Extensor strengthening allows a better balance for the joints and the antagonist muscles involved.

Fourth Week to Sixth Week

1. *Physical aspects*

- The intensity of the effort must not exceed 50 and 70 % of the climber's on-sight maximal level (ML).
- The training can't exceed three sessions a week.
- The recovery time between two sessions is 48 h minimum.
- The duration of the sessions is 01:30 maximum.

2. *Other aspects*

- Throughout that period, the climber can gradually alternate open hand postures and crimp grips, insisting on open hand postures. Crimp grip has to be progressive (sixth week) and starts with slope grip (fourth and fifth week).
- One-finger pocket or two-finger postures are still banned.
- Tactics and techniques are the main objectives of these sessions.

- What matters is not the number of passages achieved but the way the climber feels as he is seizing the holds. Consequently, the priority of the session is the climber's gradual return to practice.

3. *Complementary training*

- Throughout that period, swimming practice remains a good thing to do.
- The climber can complete his training with electrostimulation sessions – muscle strengthening program.

Seventh Week to Ninth Week

1. *Physical aspects*

- The intensity of the effort must not exceed 70 and 90 % of the climber's on-sight maximal level (ML).
- The training can't exceed four sessions a week.
- The duration of the sessions is 02:00 maximum.
- The athlete can't climb two days, one after the other.

2. *Other aspects*

- Throughout that period, the climber has to alternate open hand postures and crimp grips and gradually use two-finger grips.
- What matters is quality rather than quantity. As regards the physical aspect, muscular strengthening is one of the aims of this cycle.

3. *Complementary training*

- The climber can complete his training with electrostimulation sessions on the flexors and extensors – muscle strengthening program.

Tenth Week to Twelfth Week

1. *Physical aspects*

- The intensity of the effort must not exceed 90 and 100 % of the climber's on-sight maximal level (ML).
- The training can't exceed four sessions a week.
- The duration of the sessions is 02:30 maximum.

2. *Other aspects*

- The recommendations concerning the sessions are identical to the cycle W7–W9.

Return-to-training recap:

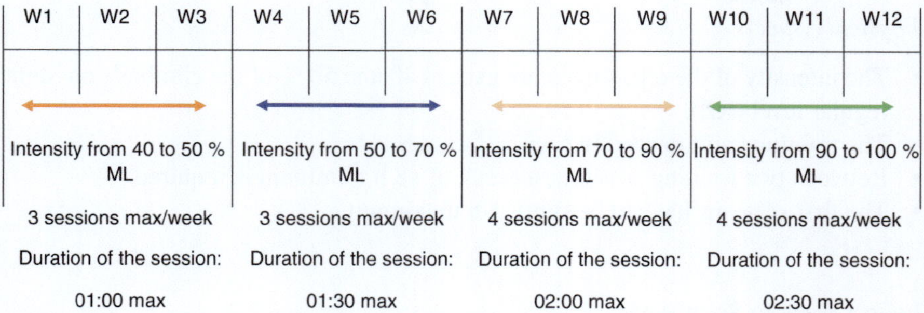

The athlete may need a couple of weeks to recover. These recovery weeks can be set between each cycle, that is to say, on the 4th week, 7th week, and/or 10th week. Of course, the next cycle will have to be postponed to the following week.

Example:

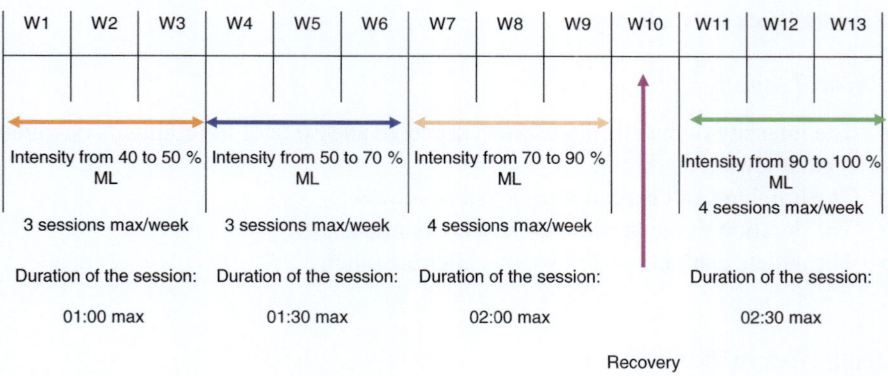

5.7.4.2 Level 2 Bone Injuries: Surgery

In this part of the book, we will only deal with the physical aspect. The recommendations concerning the other aspects are identical to those given for level 1 bone injuries.

First Week to Third Week

Physical Aspects

- The intensity of the effort must not exceed 40 and 50 % of the climber's on-sight regular level (RL).
- The training can't exceed two sessions a week.
- Between two training sessions, a break of 48 h minimum is required.
- The first sessions have to be short, 1 h maximum.

Fourth Week to Sixth Week

Physical Aspects

- The intensity of the effort must not exceed 50 and 70 % of the climber's on-sight regular level (RL).
- The training can't exceed three sessions a week.
- Between two training sessions, a break of 48 h minimum is required.
- The duration of the sessions is 01:30 maximum.

Seventh Week to Ninth Week

Physical Aspects

- The intensity of the effort must not exceed 70 and 80 % of the climber's on-sight regular level (RL).
- The training can't exceed three sessions a week.
- The duration of the sessions is 02:00 maximum.
- The athlete can't climb 2 days, one after the other.

Tenth Week to Twelfth Week

Physical Aspects

- The intensity of the effort must not exceed 80 and 90 % of the climber's on-sight regular level (RL).
- The training can't exceed four sessions a week.
- The duration of the sessions is 02:00 maximum.

Thirteenth Week to Fifteenth Week

Physical Aspects

- The intensity of the effort must not exceed 90 and 100 % of the climber's on-sight regular level (RL).
- The training can't exceed four sessions a week.
- The duration of the sessions is 02:30 maximum.

Return-to-training recap:

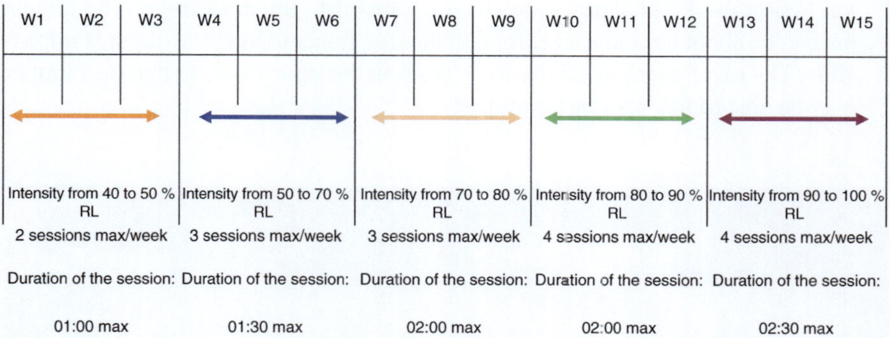

> The athlete may need a couple of weeks to recover. These recovery weeks can be set between each cycle, that is to say, on the 4th week, 7th week, and/or 10th week. Of course, the next cycle will have to be postponed to the following week.

Example:

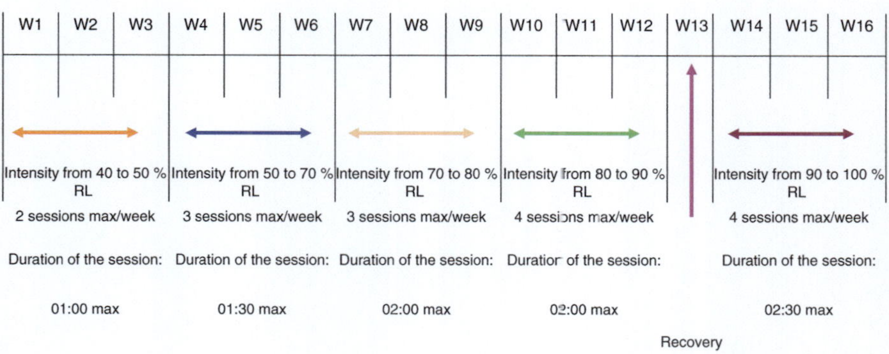

Strapping

Taping a finger is a means to relieve it for a while. Concerning this injury, it is advised to only apply that tape during the 3 weeks of the first cycle. Throughout the first 2 weeks, it is part and parcel of the session. Then during the third week, it can be removed during the warm-up but has to be applied for more difficult passages.

Thanks to that bandage, fingers are knit together and prevented from any torsion or twist. This strapping is close to a relative immobilization.

- The climber cuts off a 7-cm-long tape, as large as the P1 phalanx. First, the athlete wraps the P1 phalanx; secondly, he joints it to the next finger. The fingers have to be about the same size. Joining the ring finger to the little finger is impossible. The two finger joints have to be at about the same level, so that the climber may be able to flex his fingers properly.

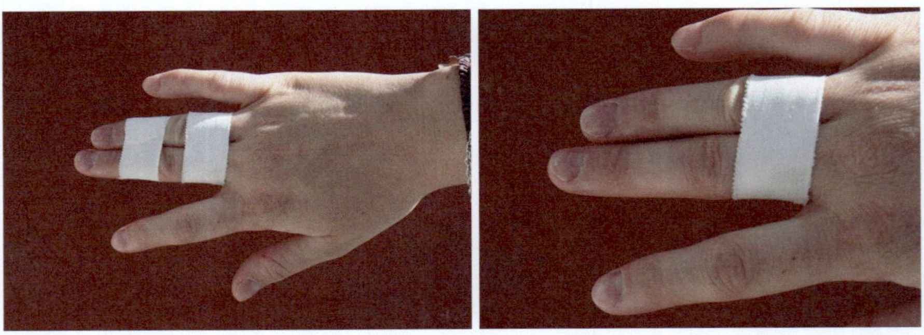

Photos 5.42 and 5.43

- The same process has to be reproduced with the P2 phalanx. P2 is wrapped around once and then joint to the next finger.

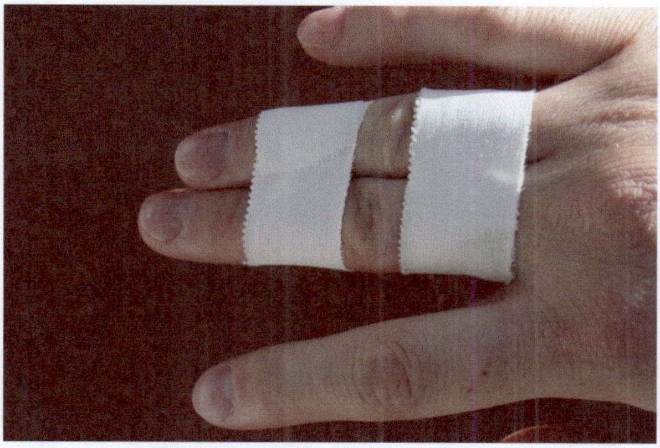

Photo 5.44

5.7.5 Ligament Injuries

5.7.5.1 Back on Track

The climber has to be extremely cautious with this kind of lesion. The slope and the width of the grips require a precise position of the fingers. The strength involved may imply a work in torsion, which is quite a strain for the joints. This is why they have to be spared by choosing carefully the grips and holds. This choice can't be made outdoors, so difficulties, routes, and cliffs have to be properly selected. A bandage is recommended once back on track. After several weeks of training, the climber will be able to apply his bandage after the warm-up, for difficulties close to 80 % of his regular level on-sight.

Warm-up is a priority especially with a rubber ball or some modeling clay. This work can be done as a reeducation during a day off.

One-finger pocket or two-finger grips, with or without the former injured finger, are to be banned. It is advised to support that finger safely with another one, at least on the holds. This recommendation is to be respected for all the fingers except for the thumb or the little finger. In case of a middle finger injury, the grip has to be done with the index, the middle finger, and the ring finger, at least.

5.7.5.2 Level 1 Ligament Injuries: Less Than 21 Days of Interruption

First Week to Third Week

1. *Physical aspects*
   - The intensity of the effort must not exceed 40 and 50 % of the climber's on-sight maximal level (ML).
   - Between two training sessions, a break of 48 h minimum is required. During the first week, it is advised not to train more than two training sessions. As regards the second and third week, the climber can go up to three sessions maximum.
   - The duration of the sessions is 01:00 maximum.
   - The recovery time between each passage has to be long.
2. *Other aspects*
   - Throughout that period, the best thing to do is to train outdoors as much as possible, choosing the adequate climbing style. Round holds are essential to help the gradual return of sensations. Alternating grips improve the relearning of climbing hand gestures. Large enough holds are favored since they enable the four long fingers to work. The former injured finger is supported by the other fingers to share equally the effort involved. The finger must not be twisted.
   - Open hand grips have to prevail. Crimp grips won't be used exhaustively by the third week.
   - The focus is on quality, insisting on techniques, tactics, and psychology.
3. *Complementary training*
   - Three times a week, electrostimulation sessions – back to training or reeducation program – can complete the training.
   - During the day, the flexors and extensors can be strengthened with a rubber ball or an elastic. Using the former injured finger to squeeze the ball, in crimp grip, is forbidden. Actually, the ball is set in the middle of the hand, and all the fingers rub it.

Throughout this period, all the finger muscles have to be warmed up properly, using a rubber ball, modeling clay, or an elastic. Hydration (before, during, and after each session) and stretching exercises (finger deep flexors, extensors, etc.) are also extremely important. After each session, it is advised to ice the finger.

The sore is always present during the first 2 weeks. The sessions have to be focused on quality. And the psychological aspect prevails to help the climber deal with the pain on his return to practice.

Fourth Week to Sixth Week

1. *Physical aspects*

   - The intensity of the effort must not exceed 50 and 70 % of the climber's on-sight maximal level (ML).
   - The training can't exceed three sessions a week.
   - The duration of the sessions is 01:30 maximum.
   - The recovery time between the different passages has to be long.

2. *Other aspects*

   - Throughout that period and according to his capabilities, the climber can alternate CG or outdoor sessions.
   - Open hand grips prevail. Nonetheless, during the session, crimp grips can be used on specific tasks. Crimp grip is practiced on large enough holds allowing the use of the slope posture.
   - The work is principally based on quality focusing on technique, tactics, and psychology.
   - The seizing has to be done with the four long fingers, so that the former injured finger might be supported by another finger.

3. *Complementary training*

   - Three times a week, electrostimulation sessions – muscle strengthening program – can complete the training.
   - During the day, flexors and extensors can be strengthened with a rubber ball or an elastic.

Seventh Week to Ninth Week

1. *Physical aspects*

   - The intensity of the effort must not exceed 70 and 90 % of the climber's on-sight maximal level (ML).
   - The training can't exceed four sessions a week.
   - The duration of the sessions is 02:00 maximum on CG. They can be longer outdoors.
   - The recovery time between the different passages has to be long.

2. *Other aspects*

   - Open hand and crimp grip postures have to be alternated during the seizing of the holds. The aim of this cycle is to regain sensations in open hand posture, as much as possible. The athlete may have to change his way of climbing to

focus on this posture. Crimp grip is practiced on large enough holds, allowing the use of the slope posture.
- The work is principally based on quality rather than quantity. Muscular strengthening is another objective of the cycle.
- Three-finger grip is given a greater importance to support the former injured finger. The athlete can start using a two-finger grip.

3. *Complementary training*

   - Three times a week, electrostimulation sessions focused on the flexors and extensors –strength or resistance program – can complete the training, according to the climber's field or objective.

Tenth Week to Twelfth Week

1. *Physical aspects*

   - The intensity of the effort must not exceed 90 and 100 % of the climber's on-sight maximal level (ML).
   - The training can't exceed four sessions a week.
   - The duration of the sessions is 02:00 maximum on CG. They can be longer outdoors.
   - The recovery time between the different passages has to be long.

2. *Other aspects*

   The recommendations concerning the session are identical to the cycle W7–W9.

Return-to-training recap:

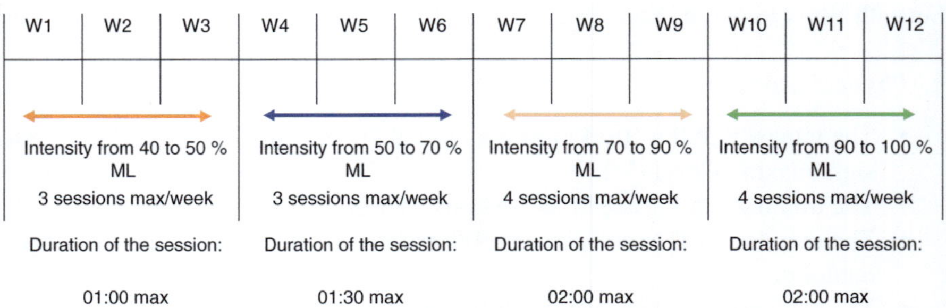

The athlete may need a couple of weeks to recover. These recovery weeks can be set between each cycle, that is to say, on the 4th week, 7th week, and/or 10th week. Of course, the next cycle will have to be postponed to the following week.

Example:

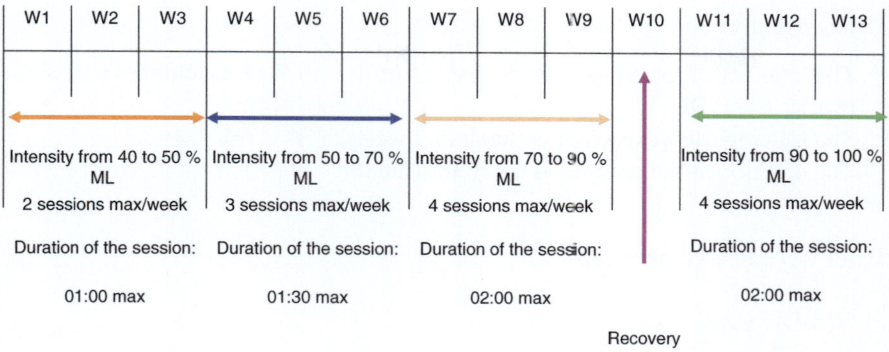

5.7.5.3 Level 2 Ligament Injuries: More Than 21 Days of Interruption

In this part of the book, we will only deal with the physical aspect. The recommendations concerning the other aspects are identical to those given for level 1 ligament injuries.

First Week to Third Week

Physical aspects

- The intensity of the effort must not exceed 30 and 40 % of the climber's on-sight regular level (RL).
- The training can't exceed two sessions a week.
- The duration of the sessions is 01:00 maximum.
- The recovery time between each passage has to be long.

Fourth Week to Sixth Week

Physical Aspects

- The intensity of the effort must not exceed 40 and 60 % of the climber's on-sight regular level (RL).
- The training can't exceed three sessions a week.
- The duration of the sessions is 01:30 maximum.
- The recovery time between each passage has to be long.

Seventh Week to Ninth Week

Physical Aspects

- The intensity of the effort must not exceed 60 and 80 % of the climber's on-sight regular level (RL).
- The training can't exceed three sessions a week.
- The duration of the sessions is 01:30 maximum.

Tenth Week to Twelfth Week

Physical Aspects

- The intensity of the effort must not exceed 80 and 90 % of the climber's on-sight regular level (RL).
- The training can't exceed four sessions a week.
- The duration of the sessions is 02:00 maximum.

Thirteenth Week to Fifteenth Week

Physical Aspects

- The intensity of the effort must not exceed 90 and 100 % of the climber's on-sight regular level (RL).
- The training can't exceed four sessions a week.
- The duration of the sessions is 02:00 maximum.

Return-to-training recap:

<table>
<tr><td>W1</td><td>W2</td><td>W3</td><td>W4</td><td>W5</td><td>W6</td><td>W7</td><td>W8</td><td>W9</td><td>W10</td><td>W11</td><td>W12</td><td>W13</td><td>W14</td><td>W15</td></tr>
<tr><td colspan="3">Intensity from 30 to 40 % RL</td><td colspan="3">Intensity from 40 to 60 % RL</td><td colspan="3">Intensity from 60 to 80 % RL</td><td colspan="3">Intensity from 80 to 90 % RL</td><td colspan="3">Intensity from 90 to 100 % RL</td></tr>
<tr><td colspan="3">2 sessions max/week</td><td colspan="3">3 sessions max/week</td><td colspan="3">3 sessions max/week</td><td colspan="3">4 sessions max/week</td><td colspan="3">4 sessions max/week</td></tr>
<tr><td colspan="3">Duration of the session: 01:00 max</td><td colspan="3">Duration of the session: 01:30 max</td><td colspan="3">Duration of the session: 01:30 max</td><td colspan="3">Duration of the session: 02:00 max</td><td colspan="3">Duration of the session: 02:00 max</td></tr>
</table>

The athlete may need a couple of weeks to recover. These recovery weeks can be set between each cycle, that is to say, on the 4th week, 7th week, and/or 10th week. Of course, the next cycle will have to be postponed to the following week.

Example:

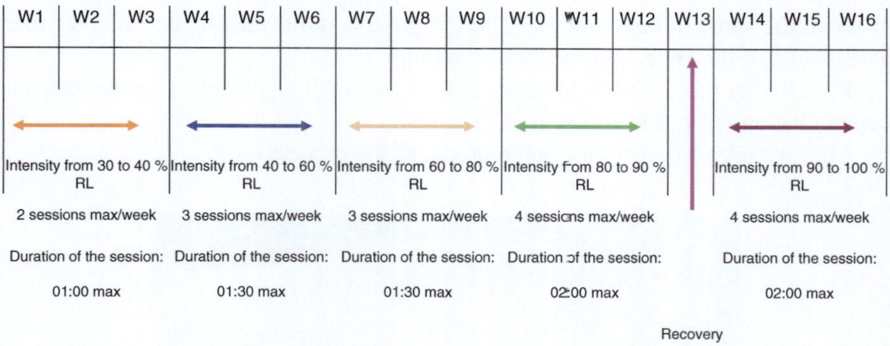

Strapping

Strapping is a means to relieve a finger for a while. As regards that lesion, it is advised to apply the tape during the 3 weeks of the first cycle. Taping is part and parcel of the first 2 weeks. On the third week, the bandage can be removed for the warm-up, but it has to be done again for more difficult passages of the session.

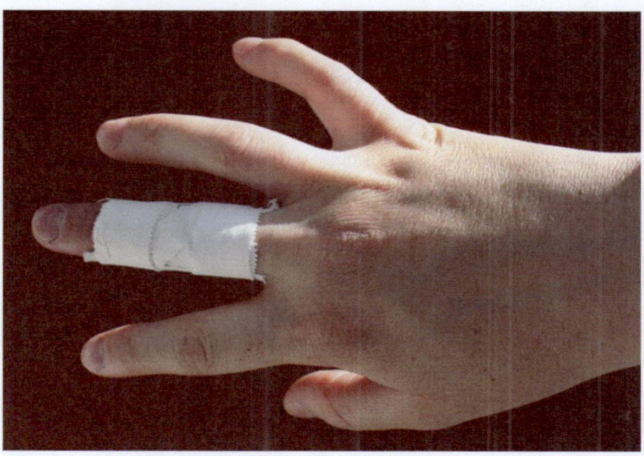

Photo 5.45

In that case of injury, the function of the tape is to maintain the finger in the right axis and to prevent it from any lateral movements. The P2 phalanx always has to work in the same direction of P1. The PIP joint is quite similar to – on a smaller level – that of the knee, and so is the strapping.

The climber takes a 8-cm-long tape, as wide as the P1 phalanx. Then he wraps it around the phalanx, not too tight.

After, he takes another tape and reproduces the process on P2.

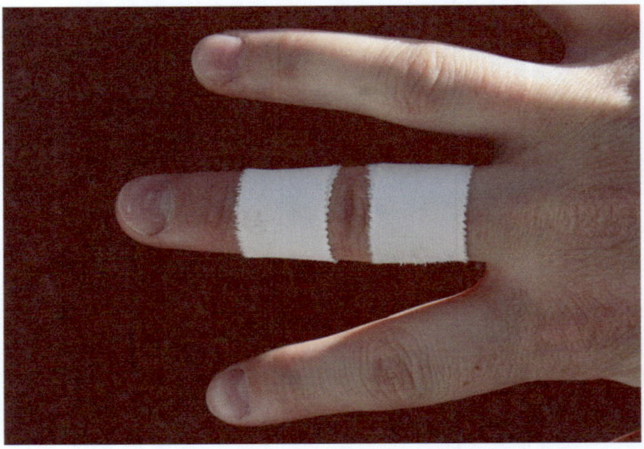

Photo 5.46

Next, the climber cuts off two bandages (length, 4 cm/width, 1 cm) and applies them across the joint. These bandages have to be tight.

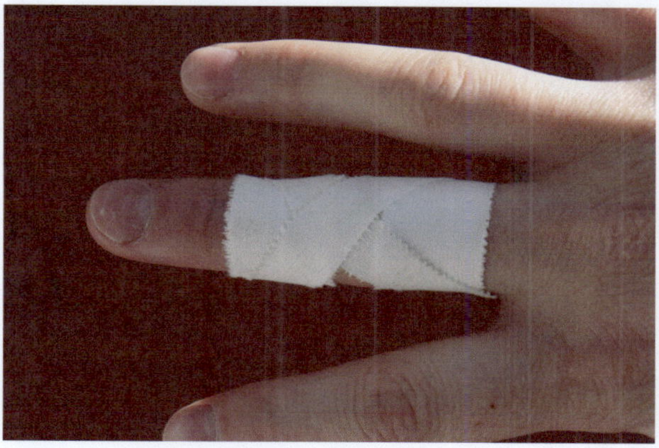

Photo 5.47

After the athlete cuts offs two small bandages (length, 3 cm/width, 1 cm). These bandages are set tightly on each side of the joint. The finger is outstretched.

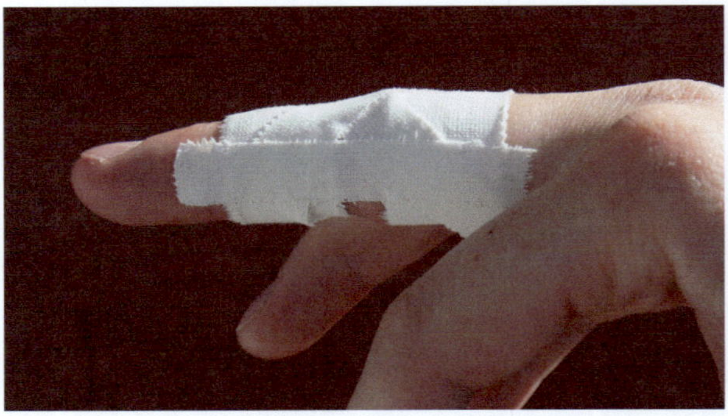

Photo 5.48

The first two operations have to be reproduced on P1 and P2 to maintain the finger properly.

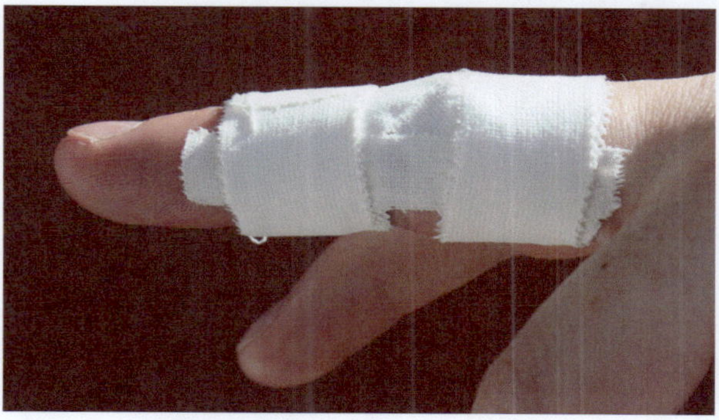

Photo 5.49

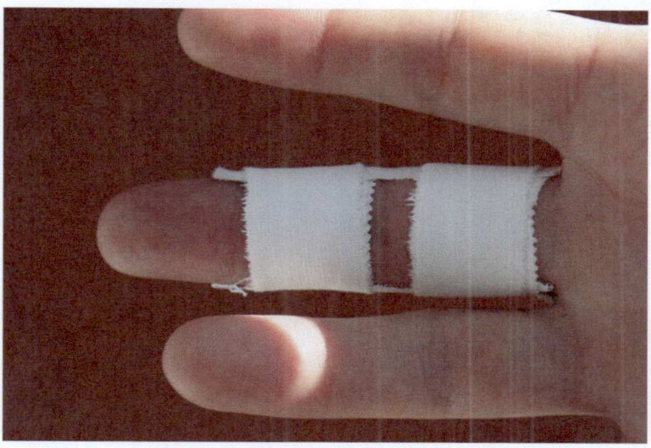

Photo 5.50

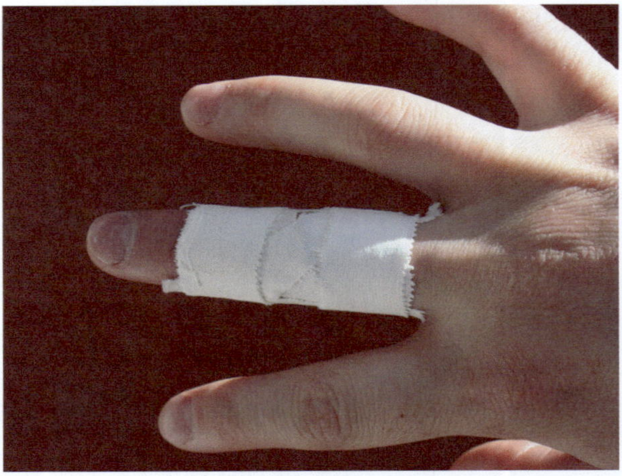

Photo 5.51

5.8 Recap Table

Injury	W1–W3	W4–W6	W7–W9	W10–W12	W13–W15
Tendon level 1	MI: 60–70 % ML	MI: 70–100 % ML			
	3 sessions max 01:30 max	4 sessions max			
		02:00 max			
Tendon level 1	MI: 50–60 % ML	MI: 60–80 % ML	MI: 80–90 % ML	MI: 90–100 % ML	
	3 sessions max	3 sessions max	4 sessions max	4 sessions max	
	01:00 max	01:30 max	02:00 max	02:00 max	
Pulley with no surgery	MI: 40–50 % RL	MI: 50–70 % RL	MI: 70–90 % RL	MI: 90–100 % RL	
	2 sessions max	3 sessions max	3 sessions max	4 sessions max	
	01:00 max	01:30 max	02:00 max	02:00 max	
Pulley after surgery	MI: 20–40 % RL	MI: 40–60 % RL	MI: 60–80 % RL	MI: 80–90 % RL	MI: 90–100 % RL
	2 sessions max	3 sessions max	3 sessions max	4 sessions max	4 sessions max
	30 min max	01:00 max	01:30 max	02:00 max	02:00 max

Injury	W1–W3	W4–W6	W7–W9	W10–W12	W13–W15
Bone level 1	MI: 40–50 % ML	MI: 50–70 % ML	MI: 70–90 % ML	MI: 90–100 % ML	
	3 sessions max	3 sessions max	4 sessions max	4 sessions max	
	01:00 max	01:30 max	02:00 max	02:30 max	
Bone level 2	MI: 40–50 % RL	MI: 50–70 % RL	MI: 70–80 % RL	MI: 80–90 % RL	MI: 90–100 % RL
	2 sessions max	3 sessions max	3 sessions max	4 sessions max	4 sessions max
	01:00 max.	01:30 max.	02:00 max.	02:00 max.	02:30 max
Ligament level 1	MI: 40–50 % ML	MI: 50–70 % ML	MI: 70–90 % ML	MI: 90–100 % ML	
	3 sessions max	3 sessions max	4 sessions max	4 sessions max	
	01:00 max	01:30 max	02:00 max	02:00 max	
Ligament level 2	MI: 30–40 % RL	MI: 40–60 % RL	MI: 60–80 % RL	MI: 80–90%RL	MI: 90–100 % RL
	2 sessions max	3 sessions max	3 sessions max	4 sessions max	4 sessions max
	01:00 max	01:30 max	01:30 max	02:00 max	02:00 max

Recovery weeks are not included in that recap table
Abbreviations: *ML* maximal level, *RL* at right regular level, *MI* maximal intensity

5.9 Interviews

Photo 5.52 Loic Gaidioz

Name: Gaidioz
Surname: Loic
Finger injuries: two tendinitis and a sprain
Rank:

- France champion in 2006
- 2006: World championship (5th)
- 2009: World championship (10th)

Best performances:

- Bouldering: 8b after work/8a at sight
- Cliff: 8a+ on sight

- *How long have you been climbing? How often do you train?*

I've been climbing since the age of 19. I train three to four times a week. On top of that, I also do cycling, ski touring, riding motor bikes, swimming, watching TV. But I also like staying quiet at home.

- *How did you get injured and how did you feel before the accident?*

I got two tendinitis on a zip down. I lost grip, all my fingers slid away, and I tried to hold myself back, once on the ring finger and the next time on the little finger. Actually, I felt perfectly well and all the conditions were great.

As for the sprain, I think it was due to tiredness, but I did not detect it before a while!

- *What about your return to training? Did you change some of your climbing habits?*

The finger lesions were not that serious, so I took some rest and everything went well. Besides, what happened was an accident, as it may occur while you're climbing, so it wasn't really my fault.

Actually, I didn't change anything in my way of climbing, apart from the fact that I have improved technically and physically and I haven't hurt myself since then, finger crossed of course!!!

- *Can you train again up to 100 % of your skills, or do you still feel a pain in the finger that has been hurt?*

I took several weeks of rest for each injury – 3–5 weeks – and since the complete recovery, everything's back to normal: I can force up to 100 % and I don't feel any pain.

- *To conclude, is there anything important you'd like to tell climbers, any advice?*

Learn to know your body! It may sound stupid, but it's important to detect and understand all the information given by our muscles, tendons, ligaments, and brains.

Above all, know one's limits…. It's better to cancel a planned session rather than get hurt and not be able to train for the next 3 months.

Photo 5.53 Loic Gaidioz in competitions

Photo 5.54 Juliette Danion

Name: Danion
Surname: Juliette
Finger injuries: pulley partial rupture
Rank:

- Europe champion in 2007
- 2007: Winner of the bouldering World Cup
- 2008: Bouldering France Champion

Best performances:

- Cliff: 8b+ after work/7c+ on sight

- *How long have you been climbing? How often do you train?*

I've been climbing since the age of 12. During the period of competitions, I trained three to four times a week, but now I climb twice a week.

- *How did you get injured and how did you feel before the accident?*

To tell the truth, since I've started competitions, I've had several injuries. Concerning the pulley partial breaking, it happened in China during the finale of the World Cup. I was about to start the circuit – at that time, the finale consisted of a circuit of six boulders – it was hot; I had warmed up, but not too much. And as usual, I had to wait a while before starting the circuit. On the first boulder, as I was crimp gripping on a small hold, I heard and felt a crack sound in my right little finger.

- *What about your return to training? Did you change some of your climbing habits?*

Going back to training was hard, because my finger was still painful. Doctors had advised me to stop climbing for 2 months. So when I started climbing again, to avoid the pain, I used the open hand posture essentially. First, it was difficult, but as time went by, I got used to it.

- *Can you train again up to 100 % of your skills, or do you still feel a pain in the finger that has been hurt?*

I can't use the crimp grip as I used to, and when I force a little, it's painful. Nonetheless, I got in the habit of using the open hand posture, so I think it won't stop me to keep the level I used to have before the injury. The only trouble is that I can't climb extreme routes, which requires the crimp grip, anymore.

- *To conclude, is there anything important you'd like to tell climbers, any advice?*

As I told you before, I've been injured several times and I've experimented a lot of cures, and for me, the best one remains applying ice on the painful lesion. It's at the same time a good means to avoid any new lesion or to treat one. After climbing, each time I feel a pain, I usually ice my fingers.

To finish, I'd like to add a few words about the frequency of training sessions. Young people always tend to climb too much. Fingers are not made for that, so the tougher the strain is, the more injured you get. Since I've reduced my climbing sessions, I've noticed that my fingers are much better and safer. If you don't give a rest to your fingers, your chances to get hurt are higher.

Photo 5.55

Photo 5.56 Rémi Samyn

Name: Samyn
Surname: Rémi
Finger injuries:

- Left hand little finger A2 pulley breaking
- Left hand volar plate breaking
- A great number of chronic tenosynovitis on the third and fourth finger of both hands

Rank:

- 2003 and 2005: University France Champion
- Junior World Championship (3rd)
- 2005: Bouldering World Cup (5th)
- 2005: Birmingham Bouldering World Cup (2nd)

Best performances:

- Bouldering: 8b after work
- Cliff: 8c after work

- *How long have you been climbing? How often do you train?*

I've been climbing since the age of 10, so I've been practicing in all fields – cliffs, bouldering, competitions, CG – for 15 years now. Since my finger problems, I've reduced the amount of climbing in training sessions, so today, I climb approximately three times a week. But on the other hand and of course to spare my fingers, I've increased nonspecific work such as fitness, jogging. Globally, I train 20–25 h a week, which makes ten sessions over 6 days.

- *How did you get injured and how did you feel before the accident?*

I broke my pulley and my volar plate when I was competing for the Bouldering France Championship, in Fontainebleau in 2008. I was pressing on a large hold, my left hand gripped in pliers, when suddenly I felt a big crack. That's when it happened! Before the competition and for 2 months and a half on, my both hands had been sore due to tenosynovitis on the third and fourth fingers. Nonetheless, I went on training, but I reduced the load of work, and I made sure to limit tough grips and harsh movements. Despite of that, my fingers had become fragile and painful, and the pressure of the competition was the last straw. My fifth finger, which had been overused to compensate during the previous 2 months, was injured.

- *What about your return to training? Did you change some of your climbing habits?*

I was back on track 8 weeks later, but in a very gradual way. Indeed, it took me 2 months before being able to deal with hard boulders again. First, I had to bandage my hands, then I tried easy boulders, and I iced my hands after each session.

I also had to change my climbing habits. I used open hand posture more often, I climbed three times a week but never more than 2 days one after the other, I did less boulders and more routes, and, finally, I insisted much more on the warm-up and I also listened to my body.

- *Can you train again up to 100 % of your skills, or do you still feel a pain in the finger that has been hurt?*

I've never managed to train up to 100 % of my skills. Each time I increase the pressure on my fingers by climbing more or strengthening my fingers with hanging work, or when I am extremely tired, I get hurt again! Changing my way of training allowed me to have no more tenosynovitis pains, but unfortunately, my little finger (double fracture) has remained painful even 1 year after the accident.

- *What about your return to training? Did you change some of your climbing habits?*

The best thing I could tell them is to be extremely careful, because once you've gotten hurt, whether a part of your body gets broken or torn, you never get back your full skills!

Another piece of advice from "an old guy": be safe and make sure to get some rest especially to spare your fingers, even if you think you don't need any. In fact, the problem is that when you are young, it's easy for your body to recover quickly, but as you're getting old, it's another story. In the end, you pay for all the mistakes you've made in your youth, but when you realize it, it's too late! It's difficult to see things in the long term, when you are 16, but it's a priority if you want to keep performing at high level. As soon as you feel a pain, you have to stop the session immediately and ice the part of the body involved. Don't hesitate to get some rest for a week, because most of the finger lesions heal up perfectly well if they are treated right away. There's nothing worse than keeping grinding and thinking that things are going to be okay despite the pain which will hopefully blur away. Finally, when you are injured, the worst thing is impatience. I know how frustrated you can feel in such periods and believe me, I've been through hardship for the last few years, but if you go too fast and don't treat your lesion properly, sooner or later you'll have to pay for it. Therefore, put up with it and be patient.

Photo 5.57

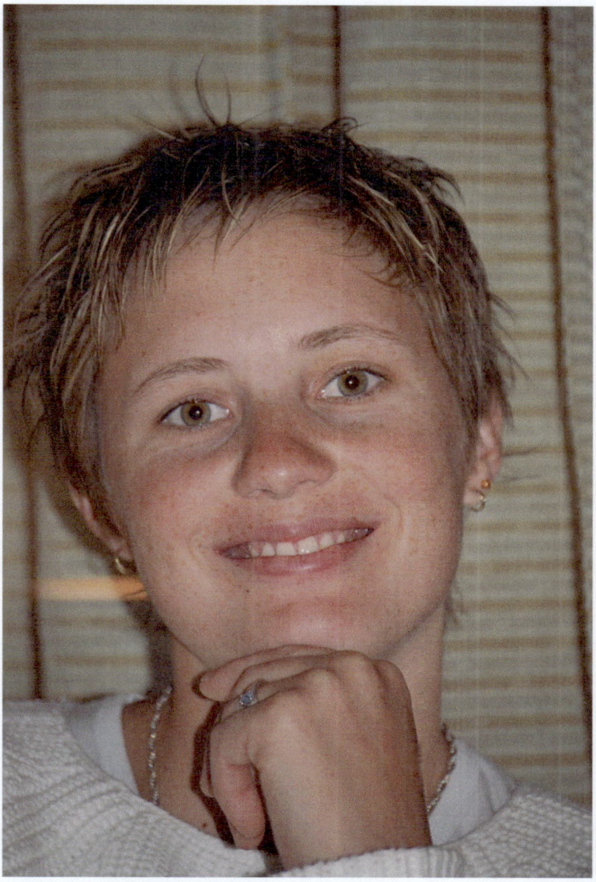

Photo 5.58 Emilie Verdier

Name: Verdier
Surname: Emilie
Finger injuries: pulley partial rupture
Rank:

- 2004 and 2005: finalist to several Bouldering World Cups
- 2004: Lecco Block European Championship (8th)
- 2005: Argentière Bouldering International Open (2nd)
- 2006: Plouha France Championship (3rd)

Best performances:

- Bouldering: 7c+ after work
- *How long have you been climbing? How often do you train?*

  I've been climbing since the age of 20. I train five times a week.

- *How did you get injured and how did you feel before the accident?*

I broke my pulley partially during a training session. I crimp gripped on a small edge as I was on a steep overhang. It was right before the beginning of the competition season and I was fit and I felt good.

- *What about your return to training? Did you change some of your climbing habits?*

I had to stop everything for 45 days before going back on track. My finger was a bit stiff at the beginning, but things went better throughout the sessions. As regards my climbing habits, I try to drink a little more during the sessions, and I stretch my fingers a little longer after climbing.

- *Can you train again up to 100 % of your skills, or do you still feel a pain in the finger that has been hurt?*

Today everything's back to normal and I don't feel any pain, because I did things step by step.

- *To conclude, is there anything important you'd like to tell climbers, any advice?*

What matters is to take care of oneself with a healthy life and to listen to one's body. When you are injured, you have to stop as long as necessary, even if it's difficult especially in case of finger injuries, because in climbing fingers are so fragile and so essential at the same time.

Photo 5.59

Photo 5.60 François Lombard

Name: Lombard
Surname: François
Finger injuries: pulley full breaking with surgery on the right ring finger
Rank:

- 1994: World Cup Winner
- 1995 and 1996: Arco Master Winner

Best performances:

- Bouldering: 8b after work
- Cliff: 8c+ after work and 8b on sight
- *How long have you been climbing? How often do you train?*

I haven't been climbing in winter for 8 years ncw, and I've hardly climbed in summer.

- *How did you get injured and how did you feel before the accident?*

I was exhausted due to a great amount of training sessions and the pain gradually appeared, so I had an infiltration done to take part in Serre Chevalier competition. A week later, as I was practicing crimp grips on the hangboard, I heard a slight crack, but no pain. Therefore, I went on climbing for 1 month nonstop, and there was still no pain. But after that, I just couldn't climb anymore; the pain was too harsh.

- *What about your return to training? Did you change some of your climbing habits?*

The return to training was gradual throughout 6 months, then I didn't change anything in my climbing habits, and I trained over my limits as I used to.

- *Can you train again up to 100 % of your skills, or do you still feel a pain in the finger that has been hurt?*

I haven't felt any pain since the surgery. Actually, the pulley seems to be even stronger than the other ones, which happened to be tricky sometimes and forced me to get some rest.

- *To conclude, is there anything important you'd like to tell climbers, any advice?*

To speak straight, though I'm not fond of the open hand posture, what I am going to say is a biased point of view.

As regards most finger injuries, a full rest is not necessary, because even after a long break, pains will always come back…. You have to go on climbing safely but bearing in mind the "no pain at all" rule. There's no use testing oneself every other day to see if things are getting better. As doctors prescribe it 45 days to help the healing!

Finally, I'm a bit skeptical concerning taping. I've used it quite a lot, and I think that it doesn't make any difference. In a way, I even think that it slows down the inner healing. Therefore, I don't use it anymore to support pulleys; I only apply it on palms and for cracks.

Photo 5.61

Photo 5.62 Mélanie Son

Name: Son
Surname: Mélanie
Finger injuries: tendinitis and finger sprain
Rank:

- 2003: Colmiane Master (1st)
- 2004: World Cup (5th)
- 2004 and 2005: Arco Master (1st)
- 2005 and 2006: France Vice Champion

Best performances:

- Bouldering: 7c after work
- Cliff: 8a+ after work and 7c on sight
- *How long have you been climbing? How often do you train?*

I've been climbing since the age of 13. I train five to six times a week.

- *How did you get injured and how did you feel before the accident?*

I got hurt as I was forcing to seize a far away hold in a tricky open hand two-finger posture. At that time, I was training everyday, and I lived an unhealthy life, I had a bad food diet, and I smoked.

- *What about your return to training? Did you change some of your climbing habits?*

I stopped climbing for 5 days, I bandaged my finger and I stopped using the open hand posture, and I went on climbing thinking that things will be okay. Unfortunately, I also had an elbow tendinitis and another one on the long biceps. Eventually, I had to stop for several months and follow numerous physiotherapy sessions.

- *Can you train again up to 100 % of your skills, or do you still feel a pain in the finger that has been hurt?*

Of course I can. Besides, I stopped practicing for nearly 1 year and I went back on track slowly. So now everything is back to normal.

- *To conclude, is there anything important you'd like to tell climbers, any advice?*

The first thing I would suggest is to stop immediately as soon as you feel a pain. It may sound like a silly remark, but it's not always that obvious.

Then don't hesitate to see a doctor! Your way of life is also important, and you have to pay attention to your food diet and drink water.

And the last but not the least, listen to your body and get some rest!

Photo 5.63

Photo 5.64

Chapter 6
Assessment and Safety

Photo 6.1

S. Gnecchi, F. Moutet, *Hand and Finger Injuries in Rock Climbers*, Sports and Traumatology, DOI 10.1007/978-3-319-16790-9_6

6.1 Assessment

More and more young people practice climbing at a higher and higher level. Trying to reach excellence implies a rise in the number of training sessions and in the duration of these sessions, which consequently leads to an increase in injuries. Most of the common pathologies are found with high-level climbing. Indeed, finger structures are extremely strained at that level. In a way, just like highly skilled climbers, the common climber also wants to be a good performer, but contrary to the former one, the latter does not follow strict rules, which could nonetheless help him to reduce risks as regards possible lesions.

The ordinary climber does not usually respect climbing basics such as good stretching exercises, proper hydration, or the simple fact of listening to his body.

Most of the time, the climber doesn't pay attention to the pain and prefers carrying on his practice rather than stopping it. He makes his own diagnosis, which often leads to a bad treatment of the lesion. This behavior seems to result in a rise in recurrences and/or chronic pains. In some cases, climbers accumulate the amount of mistreated injuries.

Photo 6.2

6.2 Warm-Up

Most of the time, warm-up is neglected and regarded as a waste of time. It is rare to see climbers warming up properly and gradually. Whatever sport one may practice, the first thing to do is of course to warm up. And fingers can't be neglected. According to one's climbing level, even the easy parts of a route may be a strain for the different muscles and joints involved. This is why the entire body and the hand of course have to be warmed up properly.

To assure a safe climbing, it is important to start any session with a gradual body preparation. Muscles and joints need a certain time to adjust to climbing specificities. Once again, let us remember that fingers have not been first conceived to climb, so they have to be well prepared. Warm-up is not only a physical preparation but also a psychological one.

The top priority is to have one's mind-set on the ongoing activity. Some of the lesions depicted in this book resulted from a lack of concentration. For instance, a loss of focus during the tight seizing of the hold on a dynamic movement may end up into a twisted finger. Therefore, it is important to warm up one's fingers before starting the session.

From a physical point of view, athletes' intensity and performance are better after an adequate and gradual warm-up. Muscles and tendons reach a rest temperature of 36°C; it rises to 39°C in full activity. At that temperature, reactivity from a nerve impulse is faster. Therefore, the risk to get hurt is reduced, thanks to a sharper concentration and a better tendon or muscle elasticity. All these factors benefit the athlete and lower the risks of injuries.

Warm-up has to be adjusted to the athlete's climbing specificities and the aim of the session. A session based on pure strength has to be longer and more particular, especially as regards fingers. In cold weather, warm-up needs to be longer, and the climber has to wear gloves after, to keep his hands warm. A sudden fall in the temperature of the fingers can be avoided by the use of a finger warmer.

The time of the day, when the climber practices, is also to be considered. On early morning, warm-up has to be longer. Indeed, after a long night, the body – fingers included – is still numb and consequently not ready to react properly or make any effort.

This book deals with hand and finger specificities, so the athlete will find some warm-up exercises below.

6.2.1 Warm-Up: First Part

The first exercises are done without touching any hold.

1. Hold your fists tight, and then open up your hand for 50 times.

Photos 6.3 and 6.4

2. Bend on your knees and rotate your hands on the ground for 3 min.

Photos 6.5 and 6.6

3. Fold up and fold back your fingers for 30 s.

Photos 6.7 and 6.8

6.2.2 Warm-Up: Second Part

Passive stretching exercises complete this specific warm-up. Stretching exercises: Short stretching movements, from 5 to 10 s, have to be done. What matters is not working on suppleness but awakening the body. Coupled with basic climbing movements, they prepare the body to the effort.

Fingers can also be warmed up by using a rubber ball or some modeling clay.

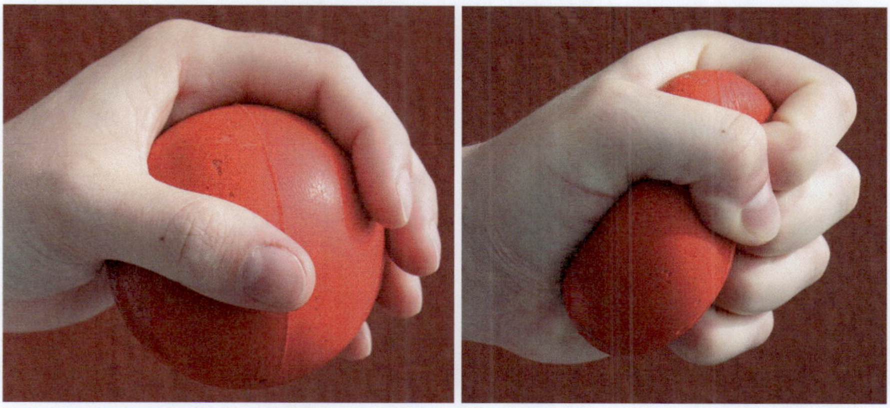

Photos 6.9 and 6.10

6.2.3 Warm-Up: Third Part

After these series of exercises, the climber gradually seizes holds on a hangboard. He can start with suspensions using his two hands to seize large holds and then try on smaller and various grips such as two-finger pockets. The athlete ends up his preparation by hanging on with only one hand. For instance, he can remain suspended for 5 s and get a 15 s break.

Photo 6.11

6.2.4 Warm-Up: Fourth Part

The athlete does traverses using all the hangboard holds. He also gradually stretches all the different muscles, alternating open hand and crimp grip posture.

Finally, he can make several attempts, lower to his top level. As the session unfolds, the movements get more and more specific and complex.

The general and specific warm-up has to last about 20 min. The body is prepared to make a top effort after 1 h of warm-up.

Photo 6.12

6.3 Stretching Exercises

Whether in climbing gyms or on cliffs, it is quite usual to see climbers practice intensively without doing any stretching exercises before or after. Nonetheless, some of these climbers will make a few climbing movements, which actually will do nothing but stretch the forearms, rather than stretch the fingers for real.

Despite that, 37 % out of the injured climbers confess that their lesions may have come from that fact. And 67 % out of the other climbers stretch while they are climbing. After a lesion, most of them had to get into the habit of looking after their fingers by stretching them after each session. These habits concerning a healthy body should be common knowledge, and the athletes should not wait to be injured to finally become aware of that phenomenon.

6.3.1 Stretching Exercises: What For?

Stretching one's body is an important part of the training.

Before any effort, stretching exercises are a means to warm up muscles, tendons, and joints and consequently to prepare the body to a physical effort.

After any effort, these exercises help promote recovery quickly by evacuating the waste and the effects of tiredness. They also allow the restoration of the muscle's initial size.

Between two efforts, stretching exercises make the recovery better and consequently the preparation for the next climbing attempt more efficient.

The function of stretching sessions is to maintain muscle and joints' precision and quality. Doing stretching exercises regularly prevents the athlete from losing flexibility throughout the years.

The ultimate goal of stretching exercises is to accelerate the healing of a lesion by reducing joint and muscle stiffness, after stopping climbing or after an immobilization.

6.3.2 Stretching Basics

Stretching exercises have to be part of each training session. The exercises may differ according to the athlete's capacities.

Several safety recommendations have to be followed.

- Stretching exercises *must not be done unprepared*. The athlete's muscles and joints have to be warmed up first.
- Muscles have to be tensed *smoothly*. No pain must be felt. Actually, the pain reveals the limit which must not be over-crossed. When a sportsman can't do an exercise properly, he has to adjust it or choose another one.
- A limb has to be stretched *slowly and gradually*, with no dash. Gathering impetus to be more efficient on the muscle is a wrong idea and might cause an injury.

- The stretching posture has to be maintained for about 15 s to be optimal. It is the requested time to allow the muscle to regain its original size little by little.

Breathing is the key word during stretching exercises. Throughout the exercises, the athlete has to breathe regularly. He has to breathe out while stretching and he must not stop breathing while he is holding the position.

Throughout the exercises, it is important to stay focused and also to listen to one's body to feel the muscles' gradual tension.

Long passive stretching exercises improve the muscles' recovery at the end of the training session. These kinds of exercises reinforce flexibility, but they don't prepare the muscles to an immediate effort. Therefore, it is forbidden to do these sorts of exercises before an important effort.

6.3.3 Finger Flexors

For sure, these muscles are the most involved in climbing, as characterized by the presence of acidosis in the forearms due to the tight seizing of the holds. Therefore, these muscles have to regain their initial size. The fact that climbers' hands look like those of monkeys is due to the fact that they never stretch their muscles, which eventually get shorter.

An ordinary person is supposed to be able to straight out his hand with an angle of 90°, without using the other hand.

6.3.3.1 Stretching Exercise 1

The sportsman bends on his knees with his hand on the ground and fingers directed to the knees. The hand and the palm have to be in contact to the surface. The elbows are tensed. To maintain the hands in contact to the ground, the climber may have to bend forward.

Photo 6.13

Holding his hand flat on the ground and his elbows tensed, the climber goes backward slowly directing his back and bedpan toward his feet. As soon as the muscle is tensed, he holds the position for 15–30 s with no dash and repeats the exercise six times.

Photo 6.14

6.3.3.2 Stretching Exercise 2

Standing, the climber rests his fingers on the other hand. Helping himself with the thumb, which is stuck to the palm, the climber tenses the forearm muscle, stretching out the wrist and holding the fingers backward. He holds the position for 15–30 s and repeats the exercise six times.

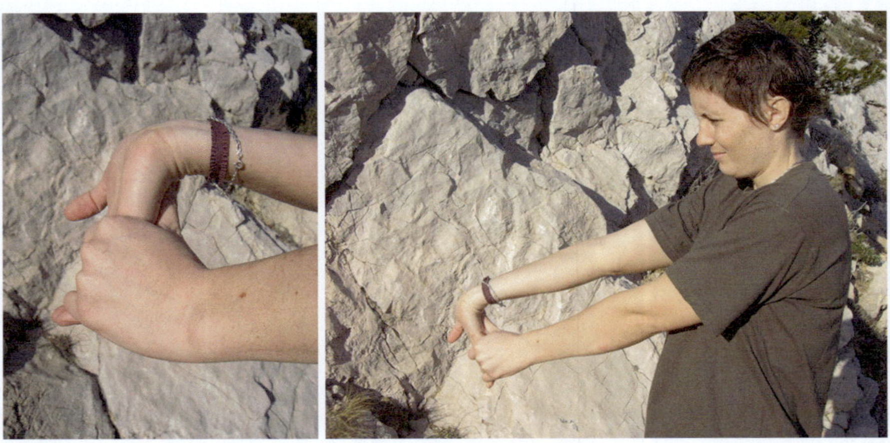

Photos 6.15 and 6.16

6.3.3.3 Stretching Exercise 3

Standing, the climber puts the tips of his fingers on a table so that he can tense his arms. He flexes his muscles by pressing his wrists slightly and downward. He holds the position for 15–30 s and repeats the exercise six times.

Photo 6.17

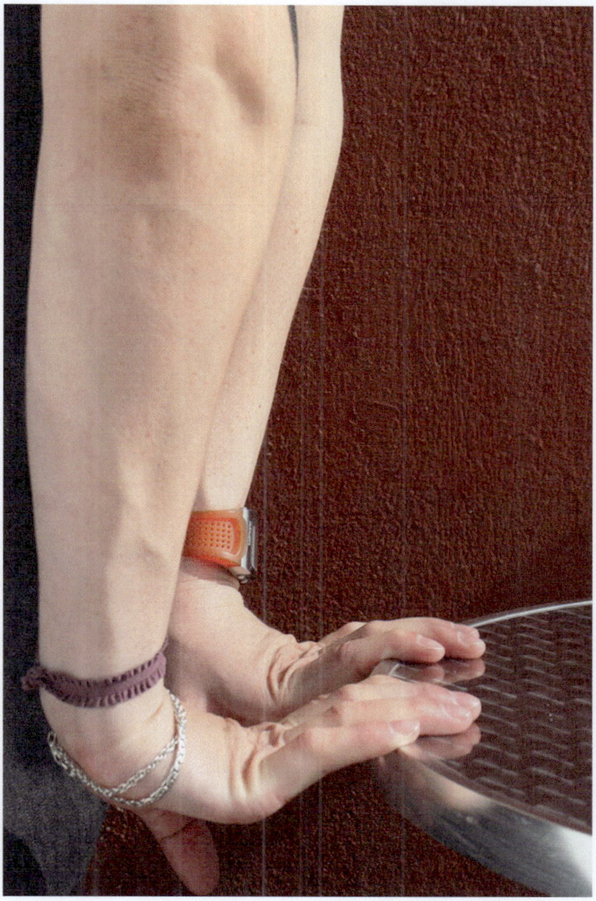

6.3.4 *Finger: Common Extensors*

The use of this muscle is hardly known by climbers, and there stands the hitch. This is why it can't be underestimated. Just like flexors, an ordinary person has to be able to stretch out his fingers, when his wrist is flexed to 90 %.

During the exercise, the tightening of the muscle is felt on the posterior side of the forearm.

Photo 6.18

6.3.4.1 Stretching Exercise 1

Standing or sitting, the climber has to fold the elbow, the wrist, and the fingers. He uses the second hand as a support to the first one.

The aim is to stretch the elbow and flex the wrist as much as possible. He holds the position for 15–30 s and repeats the exercise six times.

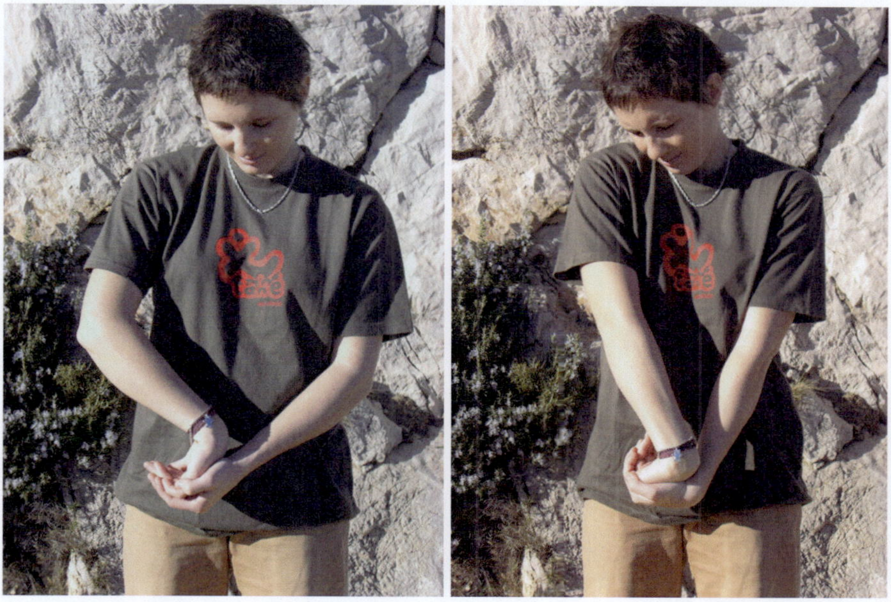

Photos 6.19 and 6.20

6.3.4.2 Stretching Exercise 2

Standing or sitting, the climber has to join all his fingers – thumb included – in a straight posture. With the other hand, he flexes the wrist. The aim is to extend the elbow. He holds the position for 15–30 s and repeats the exercise six times.

Photos 6.21 and 6.22

6.3.5 Interosseous Muscles

Most of the climbers don't know anything about these muscles. Nonetheless, they play a major part in finger coordination and joint stability. If these muscles are too short and tight, a dysfunction in the hand may appear. It is highly recommended to pay particular attention to those little muscles.

The tension has to be felt in the palm of the hand.

6.3.5.1 Stretching Exercise 1

To do this exercise properly, the forearm and the palm lean on a straight surface such as a table. It is advised to keep the fingers in the air. The climber has to fold the finger – around the PIP joint – he wishes to stretch. The same exercise can be done without leaning on a table.

Photo 6.23

The climber wraps his hand with the second one and seizes the folded finger to stretch it. He pulls the folded finger backward flexing the metacarpal joint. He holds the position for 15 to seconds and repeats the exercise six times.

Photo 6.24

6.3.5.2 Stretching Exercise 2

Do the exercise mentioned above, but this time folding the four fingers of the hand.

Photo 6.25

6.3.5.3 Stretching Exercise 3

The climber has to put his palm upward and then fold the finger inwardly, just as reported in the first stretching exercise.

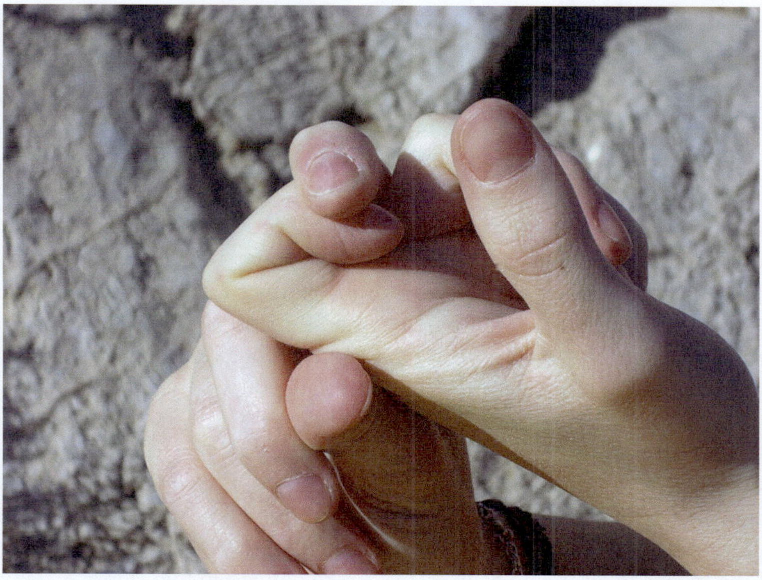

Photo 6.26

The climber uses his second hand to seize the folded finger around the third phalanx area and pulls it backward in the direction of the arm.

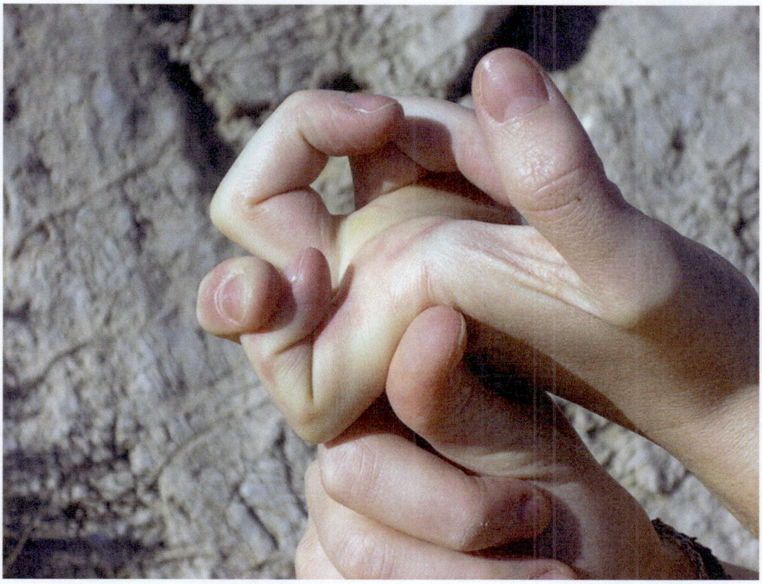

Photo 6.27

6.4 Dehydration

Dehydration is one of the causes of climbing injuries. Other factors can aggravate this problem according to the climbing sites – outdoors or indoors. In summer time, climbers are more exposed to heat, due to the reflection of the sun over the cliffs. A sunny weather with no wind is also another factor, which is hardly taken into consideration by climbers. On the other hand, according to Rubio, "when sport is practiced at high level, dehydration is higher in gyms than outdoors, because it is hotter." The ideal place should be "not too humid, cool, and well ventilated." Climbing involves dehydration problems depending on outdoor or indoor sites.

"When you are thirsty.., it's already too late…." The body is informed that water is lacking when one starts feeling thirst, it's the signal of dehydration, that is to say, water supplies are lower than waste. The mouth and the throat get dryer and less saliva is secreted.

The body is composed of 60–70 % of water. The body fights against heat by sweating. During any activity, heat comes out of the body. Therefore, to maintain the right inner temperature, the body needs to eliminate that heat. Evaporation is one of the means to eliminate heat, and it can be defined in two steps: sweating and the transformation of this sweat into steam, which is the only method to lower the body temperature.

A lack in water can lead to cramps, tendinitis, and renal calculus. Therefore, a sportsman suffering from dehydration is bound to perform badly. Indeed, 2 % of dehydration equals to 20 % loss in performance. For a person weighing 70 kg, it represents a loss of 1.4 l of water.

When people don't practice any physical activities, half of the water supply is provided by food diet. Therefore, to make up for the lack, it is necessary to drink about 1.5–2 l a day. When people do sports, the loss of water is higher, so it is advised to drink before, during, and after the activity to make up for that additional loss of water. Depending on the temperature, the need in water can be more important for climbers. Therefore, it is a priority to drink before feeling the thirst.

Unfortunately, the body can't store all the water, so it is completely useless to drink a great amount of liquid before any effort, because that water won't make up for the later loss. *On the other hand, it is important to be well hydrated before any effort in order to perform well but also to avoid any muscle or tendon lesions.*

During the effort, the athlete has to drink regularly every 10–15 min, but it is forbidden to drink 1 l of water at one go.

After the effort, it is necessary to keep on drinking in order to make up for the previous loss of water, though it is difficult to evaluate the proper quantity lacking. The quantity depends on the individual but also on the intensity of the exercises of the training session. If the session has been long and quite difficult, the loss of water can equal to 0.3 l up to 2 l. Checking the color of the urine after the effort is a good means to see whether someone is well hydrated or not. A dark yellow urine with a pungent smell shows that the body finds it difficult to evacuate the waste and is suffering from dehydration.

To find out the quantity of water loss and to make up for this loss, it is possible to do the following test but only once in your life! Check your weight before a training session and don't drink or urinate throughout the session. Then check your weight again after the session, before drinking again. Bearing in mind that the loss of weight is mainly water and that one liter equals to 1 kg, you will find out the quantity of water necessary for your future sessions as long as they are similar in intensity, duration, temperature, and humidity and you wear the same clothes. Then it will be easier to adjust the quantity of water according to all these different factors which have an impact on water loss.

The water lost by the body is full of mineral salts. Consequently, it is advised to drink alkaline water, naturally salted to make up for the loss. Fizzy water helps the recovery after the effort.

6.4.1 Causes of Dehydration Injuries

When tendons are well hydrated, they slide correctly in their sheaths. When the body is dehydrated, lubrication is prevented and frictions appear leading to tendon inflammations. During the effort, toxins are developed, and due to a bad hydration, tendons are at risk and injuries may happen.

To conclude, it is important to say that "it is a priority to learn how to drink before feeling the pangs of the thirst." Even if it is impossible to drink during a climbing attempt, it is a top priority to drink between each route or each boulder especially during competitions. Concerning long efforts – 40/50 min – such as the attempt of an on-site route, it is highly advised to have a good hydration before and after the attempt. In gyms or outdoors, it is usual to see climbers without any bottle of water. Remember that hydration is the key word even if one doesn't feel the pangs of thirst.

Climbing is touched by dehydration due to the intensity of the effort requested but also to the environment – indoors or outdoors.

Photo 6.28

6.5 Food Diet

Climbers' food diet is always a tricky question to debate. Nonetheless, a healthy food diet is vital for high-level climbers. A few reports have revealed malnutrition cases in climbers, whose fat rate is sometimes quite alarming. Apart from the climbers' health which is at risk, fingers can also suffer from an unbalanced food diet.

The climber is always obsessed by the right balance between weight and strength, an important factor as regards climbing. It is easy to picture the difficulties of a corpulent person trying to climb at high level. The easy answer to that kind of problem is to think that "losing weight will make the climbing easier and the performance better." Between the right weight during the competition period and the risks of diseases due to malnutrition, the limit is easily crossed. And the climber may enter a vicious circle leading him to a dangerous behavior as regards food. Anorexia and bulimia are quite usual in the climbing field; the climber's loss of weight allows him to perform better, and this spiral can urge him to carry on too strict diets.

The basics as regards food are to make up for the loss of calories during the effort. When these principles are not respected, performances can be diminished, and in more serious cases, diseases such as bulimia or anorexia can appear. In order to reach the right balance between weight and strength and to perform well, the climber reduces his calorie consumption to the extreme and avoids some essential nutriments. Likewise, lipids are completely banned though they are quite necessary for climbers. Indeed, it has been reported that sportsmen who consume little lipids are more likely to be injured.

The sportsman's diet fuels the effort. The chemical energy produced by food is transformed into mechanical, electric, thermal, and chemical energy. The mechanical energy produced by the muscles is the source of any daily physical or sports activity.

Food is made up of different nutrients such as carbohydrates, proteins, lipids, water, mineral salts, trace elements, and vitamins.

6.5.1 Macronutrient Classification: Carbohydrates, Proteins, and Lipids

To function properly, the body needs a specific amount of macronutrients. Each of this macronutrient plays a particular part in body functioning and can't be replaced.

Below is the daily requested amount of these energy needs.

- Carbohydrates: 60–65 %
- Proteins: 15 %
- Lipids: 20–25 %

A sportsman is supposed to have 3,000 or 4,000 cal a day, but the quantity may change depending on the physical activity.

A diet with too many carbohydrates, >65 % a day, or fat free, <15 % a day, increases inflammatory factors and consequently the risks of injuries.

During an exercise, the first energy nutrient to be used by the body is blood carbohydrates – stored as glycogen in the liver and muscles. Then fat supplies stored in adipose tissues and lastly proteins are eventually used too.

6.5.2 Carbohydrates

Most of the carbohydrates come from fruit, cereals, and vegetables. Glycogen is then transformed into carbohydrates to circulate in the blood. Carbohydrates once turned into glycogen – a mixture of several carbohydrates molecules – can be stored away by the body for the next efforts.

During any physical activity, carbohydrates transformed into glycogen fuel muscles for the main part. They are indispensable for any training session. On the average, the supplies equal to 1,200–2,000 kcal which is equivalent to an intensive training session of 1:30. One gram of carbohydrates equals approximately to 5 kcal.

Carbohydrates can be divided into two categories: the simple carbohydrates and the complex ones. Sugars are classified according to the glycemic index (GI). Simple carbohydrates tend to have the "-ose" ending such as fructose, saccharose, or lactose. They are also known as "fast sugars" since they are quickly assimilated by the body. They can't be used as supplies before an exercise, but they can fuel the body during the effort. Fast sugars have a high glycemic index, up to 70 %.

On the other hand, complex carbohydrates can be stored away. They are slowly assimilated by the body, which accounts for their name, slow sugars. Most of the time, they come from seeds and can be found in bread, cereals, pasta, rice, potatoes, or vegetables. The glycemic index is quite low and inferior to 55. They provide energy gradually.

When people practice a sport, it is better to use slow sugars. Muscular glycogen is used throughout any physical activity, and the supplies can't be restored at the same time. Therefore, when they are running short, the body starts to synthesize carbohydrates. If the sportsman remains in that state for too long, his body feels like people with food problems; the recovery time is longer and the muscles are weakened by the body. This loss in glycogens increases the risks of lesions.

In climbing, the effort is for the main part short and intensive, so muscles are fueled by carbohydrates. When the climber works outdoors for the day, the supplies are not sufficient. Therefore, it is a priority for him to feed himself properly throughout the day to maintain a right level of carbohydrates.

6.5.3 Proteins

Proteins can be animal or vegetal, so they can be found in meat, eggs, milk, cereals, soya, and vegetables. Unlike carbohydrates and lipids, proteins are not a simple source of energy. In fact, their function is to create body cells such as those in muscles. The more sports one practices, the more proteins one needs.

Proteins are used by the whole body, but they are not a real source of energy. Actually, they are essential to muscle development.

One gram of proteins equals approximately to 4 kcal.

6.5.4 Lipids (Fats)

Most of the lipids are found in fatty and oily food such as meat, margarine, and eggs. Though fat tends to have a bad image, it is nonetheless extremely important for the body. Two types of fats exist – the saturated and the unsaturated fats – and they have a different function. Sportsmen are supposed to have two thirds of unsaturated fats and one third of saturated fats. Unsaturated fats help the oxygen flowing. It has been reported that omega-3 – unsaturated fatty acid – improves blood circulation and helps reduce inflammation. Omega-3 is currently found in seaweeds and fish fat. It lowers the risks of lesions. Athletes who don't eat much fat are more likely to get hurt.

This is why it is important for the athlete to learn how to sort fats rather than do away with all of them. For instance, there is 75 % of fats in pork but only 15 % in turkey. Besides, vegetal fats have to be included in the food diet to limit animal fats.

One gram of lipids equals approximately to 9 kcal.

Fats ease vitamin circulation.

6.5.5 Extreme Malnutrition Symptoms

Depression, tiredness, and less good performances are the usual signs of repeated nutritional deficiencies. A loss of muscles can be noticed in male climbers, while female climbers may have menstruation problems. Moreover, climbers are more subjected to infections.

Though getting good results prevails, the athlete's good health is what really matters. Therefore, coaches and relatives play a main part in that field and have to pay attention to the athlete's food diet.

A classification allows detecting any weight problems according to the athlete's height. The BMI – body mass index – goes as follow: BMI = mass/height2 – mass in kg/height in meters.

BMI – kg/m^2	Results
Less than 15	Exhaustion – lack of nourishment
From 15 to 18.5	Thinness
From 18.5 to 25	Normal corpulence
From 25 to 30	Overweight
From 30 to 35	Moderate obesity
From 35 to 40	Severe obesity
More than 40	Massive obesity

When the athlete is commonly checked on, food problems can be avoided and his health saved.

6.5.6 *Vitamins*

These substances are not a source of energy, but they are essential to the good functioning of the body. They are provided by food and can't be synthesized by the body. Vitamins are divided in two categories: fat-soluble vitamins which are absorbed at the same time than fats and stored away and water-soluble ones which can't be indefinitely stored and are eliminated through urine when they are in surplus.

Let's say a few basic words about vitamins involved in sports and those leading to injuries.

Vitamin A is anti-fatigue and anti-infection. It is essentially found in fish, meat, milk, yolk, or vegetables.

Vitamin D helps calcium and phosphorus assimilation. In case of shortage, it may lead to rickets, osteoporosis, and osteomalacia. It can be found mainly in fish, cereals, or butter.

Vitamin E nourishes the muscles and skin. Its deficiency can lead to stiffness of tendon sheaths. It can be found in cereals, vegetal oils, butter, yolk, or greens.

Vitamin B2 plays a great part in energy metabolism and in skin nourishment. It is one of the most common vitamins which can be found in all kinds of food. Therefore, its deficiency is quite rare.

Vitamin B5 interferes in nutrient metabolism – carbohydrates, lipids, and proteins. This vitamin can be found in animal or vegetal food.

Vitamin B8 is part of the process to synthesize fatty acids and to produce energy with Carbohydrates. Its deficiency though rare leads to tiredness, nauseas, anorexia, and muscle pains. This vitamin can be found in yolk, soya, lentils, cereals, or fish.

Vitamin B12 interferes in carbohydrates, lipid, and phosphorus metabolism. It helps in protein synthesis and the assimilation of the former by the tissues. This vitamin can be found in meat, milk, cereals, yolk, vegetables, or fruit.

6.5.7 *Mineral Salts*

Just like the passage dealing with vitamins, the following information will focus on mineral salt effects on sports and injuries.

Calcium is the main constituent of the bone which keeps renewing. Calcium plays a great part in nerve impulse conduction and in muscle contraction. A deficiency in calcium may lead to decalcification and fragility of the bone. Calcium can be found in all kinds of food, but essentially in hard cheese, milk, or dry fruit.

Phosphorus just like calcium is a main constituent of the bone structure. It also interferes in nerve and muscle function. Calcium can be found in all kinds of food; this is why calcium deficiency is rare. Cheese, yolk, and milk are rich in phosphorus.

Potassium plays a part in nerve impulse and muscle contraction. A lack in potassium may lead to cramps. Vegetables, dry fruit, meat, fish, and bananas are rich in potassium.

Sodium plays a major part in muscle contraction and reactivity. Sodium deficiency can be extremely serious and results from a loss not compensated during the effort. Sodium can be found in salt, cooked pork meat, milk, yolk, fish, and meat.

Magnesium produces energy and transmits nerve impulse. The needs in magnesium are higher when a sportsman practices at high level. A shortage in magnesium results in muscle failure, cramps, a lack of go, and tiredness. Whole cereals, dry fruit, seashells, rice, bananas, and wholemeal bread are rich in magnesium.

Photo 6.29

6.6 Extensors

6.6.1 Common Finger Extensor Work

6.6.1.1 Theoretical Analysis

Since the 1980s, climbing has really changed. Practice has evolved, facilities have been improved, and training has become more specific. Nonetheless, climbing competitions are still quite new, and all the impacts of this intensive sport at high level are not well known.

To improve climbing training, it is necessary to consider previous researches, to make experiments, but also to study other sports. Indeed, each sport has its own specificity and muscles do not interfere in the same way, but nonetheless, some common basics can be compared.

Concerning climbing, movements are more and more physical, weight applied on holds is heavier and heavier, and fingers haven't been conceived for such a pressure. Moreover, good results are achieved younger and younger, and finger, wrist, and shoulder injuries are more and more numerous.

From a biomechanics point of view, one of the causes of these injuries could be a muscle unbalance between the flexors and extensors. This phenomenon is quite common for climbers. The constant repetition of a specific movement and a particular training tends to affect the flexors. On the other hand, climbing practice is quite traumatizing. Most of the time, injuries happen during an intensive effort or when fingers are strained to the limit, as may be the case during an attempt on a campus board. In climbing, flexors are particularly solicited especially when they are opposed to extensor contraction. Any muscular unbalance – agonist-antagonist – may lead to joint dysfunction and tendon injuries.

This phenomenon requires the coach's complete attention. This muscle unbalance is quite obvious for the climber who, on the one hand, strengthens his flexors intensively and abruptly while he is climbing outdoors or on specific appliances such as the hangboard or campus board but, on the other hand, reinforces his extensors only while he is climbing. Therefore, this unbalance has to be compensated by a global fitness during the general physical preparation. Unfortunately, this important stage of muscular balance is barely fulfilled and sometimes it's even overlooked.

Injuries resulting from extensors are not the only reason to focus on that point. Though working on that muscle is far from being exciting, it is more than ever necessary to take a closer look at it.

6.6.1.2 General Point of View

The body is composed of muscles which act and interreact between themselves. This fact accounts for the usual straight posture of the body while standing. A muscle always interreact with another one. When muscles work in the same direction, they are called agonist muscles, and antagonist muscles for the reverse. A natural balance between these two kinds of muscles allows the body to maintain its original posture. If this balance was not respected, the body would bend forward. The same dysfunction could happen to the hand due to muscles acting on finger joints. Indeed, some agonist muscles such as superficial and deep muscle flexors have a precise function and are directed to a close target, whereas antagonist muscles such as finger extensors act against opposition moving the joint in the opposite direction. A climber's hand is quite characteristic due to its constant flexion of the fingers also known as "the monkey posture." What can be said about this fact is that, first, after a training session, finger flexors do not regain their former position; in a word, climbers do not stretch short muscles and deep flexors well enough. Secondly, flexors are highly developed whereas extensors are overlooked. This aspect will be dealt with more precisely later on.

But to what extent are flexors useful in climbing?

To mobilize a finger joint, the interaction of antagonist and agonist muscles is necessary. The joint activation during an antagonist movement has two functions: precision and protection.

6.6.1.3 Gesture Precision

Coordination between antagonist and agonist muscles has to be extremely sharp to perform the most precise movement. Accuracy in coordination is a priority to achieve a perfect stability and make a successful attempt. When the coordination fails due to a bad timing or a miscalculated strength, the movements are not what they were supposed to be. In the case of dynamic movement commonly used in bouldering, a bad coordination between agonist and antagonist muscles prevents the climber to seize the hold tightly at the right time. In that particular case, the difference between the expert and the fresh climber lies in the way the different muscles are tensed.

Agonist muscles have to act simultaneously or supportively to create a smooth and regular movement. Antagonist muscles have to work alternatively with agonist muscles extending while the latter are flexing in order to avoid any resistance to the ongoing action.

During a cyclic movement, alternating flexion and extension is quite a problem in coordination, but solutions can be found throughout training on various climbing supports.

6.6.1.4 Joint Protection

When speed is involved, antagonist muscles play a main part. If they weren't contracted during a movement at fast speed, damages would be serious. Concerning dynamic movements, extensors are mainly important since they are supposed to maintain the joints' stability, especially when the fingers come into contact with the hold.

The joint activation and the mutual contraction of the muscles are important since it keeps the joints stable. When the joint activation is inadequate, subluxations around the wrist area can be noticed, in other words the wrist "shivers."

6.6.1.5 Performing: Extensors in Action

Likewise, soliciting extensors increases good results in competitions. Indeed, in sports, coordination can't be limited to the only muscles mobilizing a particular joint. The whole muscle chain matters to make the right contractions possible. And each part of this chain has to act together. According to Pradet's words, "The muscle chain strength is limited by its weakest link." This is why it is a priority to urge the climber to reinforce his extensors during the PP. The climber over-strengthening his flexors at the expense of his extensors will not be fully prepared. Therefore, reinforcing one's extensors to the excess is useless if a specific work has not been done on the flexors too. Any training session aiming at developing finger strength should be completed with a specific work on antagonist muscles.

Thanks to L. Vigouroux's biomechanical pattern, it is possible to determine the tendon tensions mobilizing each finger. Several muscles interfere in finger extensions. Apart from the FCE, intrinsic muscles such as lumbrical muscles and dorsal interosseous muscles are also involved. They act on the PIP and DIP joint extension (Eyler and Markee 1954).

FCE interferes in open hand and crimp grip postures, but less in the crimp grip. In open hand posture, its action is intensive. Indeed, in open hand posture, the weight involved on the middle finger for an average strength of 100 N is about 35.5 N, whereas in the crimp grip, it equals to 18 N. Of course, the strength involved is much inferior to the one related to the DCF, which equals 188.9 N; nonetheless, that fact can't be overlooked. Considering the fact that FCE has not the same characteristics with the flexors, these data represent a high percentage of its maximal strength. Moreover, extensors though antagonist get as tired as flexors (Quaine and Al 2004). In other words, in climbing, extensors get as tired as flexors as far as grips are concerned. This phenomenon shows that extensors are overused regardless of their capacities; this is why they have to be highly considered.

Other sports activities such as sailing or tennis pay particular attention to these antagonist muscles.

6.6.1.6 Antagonist Muscles' Functioning

Let's take a closer look to the way antagonist and agonist muscles work together.

- In the case of a slow and balanced movement in a still position, the antagonist muscle – the extensor – will stretch out and loosen up. Its action will be hardly noticed. This phenomenon concerns simple movements and implies a good level as regards coordination.
- In the case of a fast and a powerful movement in an awkward position, the antagonist muscle's action is much more complex. In fact, antagonist muscles will be directly involved in the action. They will be part of the motion but also of the joint support. Let's consider the case of a dynamic movement or that of a dyno. In those cases, the extensor will "open up" the climber's hand so that after the contact with the surface, the fingers can seize the hold in an isometric contraction to stabilize finger and wrist joints. Therefore, its function is quite important for controlling the movement but also for avoiding joint injuries.

So to conclude, it is important to highlight the extensor's role in climbing. Moreover, it can be said that as regards progresses and performances, the more complex the movements will be, the more solicited extensors will be.

6.6.1.7 Conclusion

During any physical preparation, a coordinated work between antagonist and agonist muscles has to be considered by the coaches. The first thing to do is to pay attention to the hand and fingers which have to be straightened up and well oriented with the forearm. This work will help to balance the strength involved in the fingers and consequently will reduce injuries. Moreover, coordination will be emphasized and gestures will be more precise, thanks to a better joint stability. Strengthening extensors will help flexors to be perfectly used especially when it is known that extensors are the weakest link in muscle structure.

Photo 6.30

6.6.2 Extensor Muscles: Strengthening Exercises

Strengthening extensor muscles can be done in several ways. The chosen exercises should always be done when a specific work is done on flexors especially on hangboard or campus board.

The first exercise requires the use of an elastic. The latter must not be too tight to allow a great number of repetitions, rather than try to unfold the fingers with an elastic too tight. The climber sits down and rests his elbow on his thigh; the arm has to remain close to the body. The fingers which need to be strengthened are wrapped up by the elastic which is maintained by the other hand. This exercise can complete a hangboard or campus board session. For instance, after a climbing attempt, the climber can do this exercise for 30 s on. During a hangboard session and a one-minute contract, he will be able to include 30 s of work on the extensors.

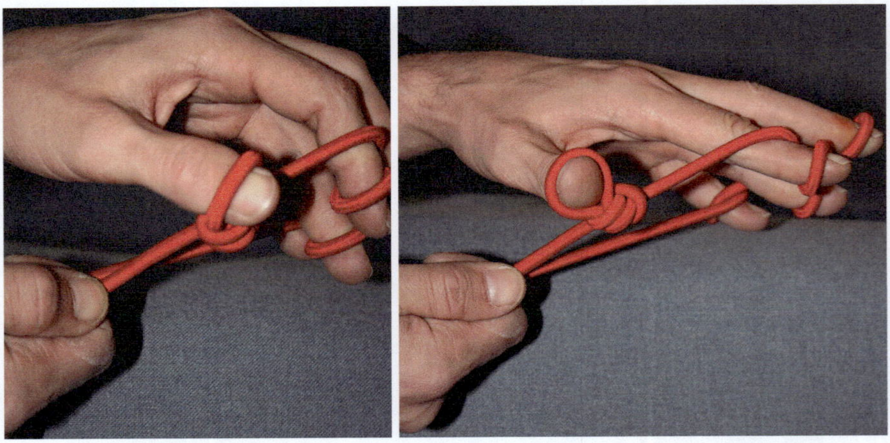

Photos 6.31 and 6.32

To practice more specifically, appliances such as the Power Finger do exist. Just as in the previous exercise, hangboard or campus board sessions can be combined to the work.

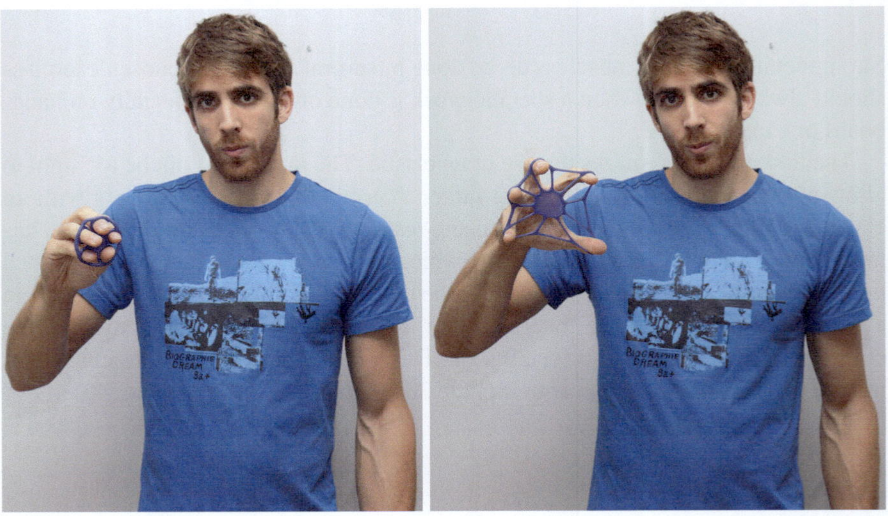

Photos 6.33 and 6.34

Electrostimulations are another way to strengthen muscles. The climber can work on extensors and flexors at the same time. Antagonist and agonist muscles are developed too.

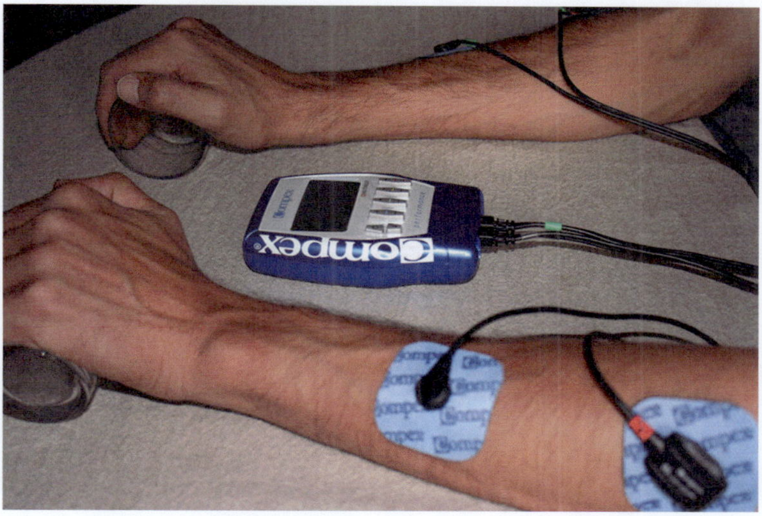

Photo 6.35

Another exercise which is much centered on the wrist and does not require a specific work on the fingers can be done with a barbell. The weight depends on the climber's capacities, but cannot be over 5 kg. The aim is to repeat the movement as many times as possible and not to get injured in case of a too heavy load.

So the climber is sitting on a chair with his forearm resting on his leg. The hand is in pronation. The exercise consists of raising and putting down the hand making sure to keep the forearm in contact with the leg.

This exercise can be done for 30 s to complete a hangboard or campus board session.

Photos 6.36 and 6.37

Photo 6.38

6.7 Electrostimulation

6.7.1 Introduction

Electrostimulation (ES) has been existing for more than 40 years. Electrostimulation training sessions have become quite common lately, and high-level athletes can't do without them anymore. This means can be used as a treatment or as a way to strengthen muscles. The latter use is quite controversial due to the different results reported by various studies. Nonetheless, the common point is that this method, combined with mechanic movements, strengthens muscles. For coaches and athletes, ES is an interesting complementary way to body fitness. The results achieved vary according to the stimulated muscles. For the ordinary climber who is running out of time, ES is a good means to train in order to progress.

A classical muscle strengthening session is quite tiring and does not contract all the muscle fibers, contrary to an ES session. Therefore, it is better to combine these two ways of training to optimize the session.

Regarding ES, a few basics have to be remembered. Concerning muscular development, the efficiency of ES depends on the number of fibers involved. Putting the electrodes on and waiting will be vain. What matters is to act on a great number of fibers insisting on the deeper ones. It is a bad idea to think that another activity can be practiced at the same time than ES. In the case of a strength program, the intensity is maximal and can hardly be endured. Back-to-training programs and reeducation programs allow the athlete to get back to practice slowly and smoothly. But the instructions have to be followed precisely.

A great number of appliances are available on the market, so they have to be chosen according to the objectives. Muscle strengthening program – strength or resistance – is available on most of the devices unlike analgesic programs or reeducation programs which are fewer though quite useful for climbers.

Programs such as muscle strengthening, reeducation, or painkiller are predetermined by retailers who record impulsion frequency or duration, contraction duration, and recovery time. New methods can assess the athlete's level of tolerance and adjust the stimulation parameters according to his physiologic characteristics.

In case of injuries, it is hard for any athlete to face a complete interruption of sport practice. ES is a good means to come back on track slowly despite the loss of muscles, the pain, and the desire to be back to the top the sooner the better. A specific work with well-adjusted stimulation parameters eases the work on a previously injured part of the body. Specific programs are usually available on ES devices designed for athlete's training. ES is commonly used in muscle strengthening or reeducation after a long immobilization due to a lesion. This training complementary method can't be overlooked.

ES can be used as a treatment to prevent muscle atrophy after a long-term interruption or to relieve pain and fight against inflammation. Electric skin neurostimulation (ESN) is used to kill pain by blocking painful nerve impulses coming from the brain. The electric current has a low intensity. The pulsations are supposed to relieve the pain. The sportsman sets the electrodes on the sore area or along a specific nerve crossing the painful area.

6.7.2 Electrode Setting

Electrodes have to be set in a precise way. According to their positions, they define the muscle to work on. This is why climbers have to set them according to the recommended patterns or as can be seen on the pictures below.

The more distant the electrodes are, the deeper muscles are stimulated.

Nonetheless, DCF muscle seems to be more difficult to stimulate due to its location.

6.7.2.1 Muscle Strengthening or Reeducation

For these programs, electrodes are placed on the forearms. When electrodes are badly positioned in that area, pins and needles may occur in the little finger or in the thumb. In that case, climbers only have to readjust them slightly.

A combined work between antagonist and agonist muscles – extensors and flexors – is advised. To work these muscles, a hard and round object, like a bottle for instance, has to be held tight in the hands while contracting. The fingers are outstretched and not positioned in crimp grip. It is advised to use the flexors rather than extensors.

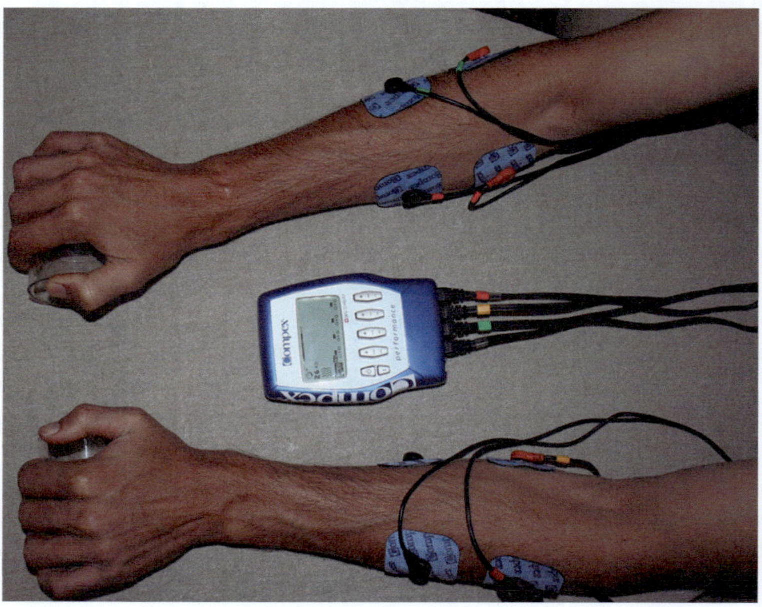

Photo 6.39

The climber can use his other hand to contract the muscle he wishes to work and after set the electrodes in the appropriate area.

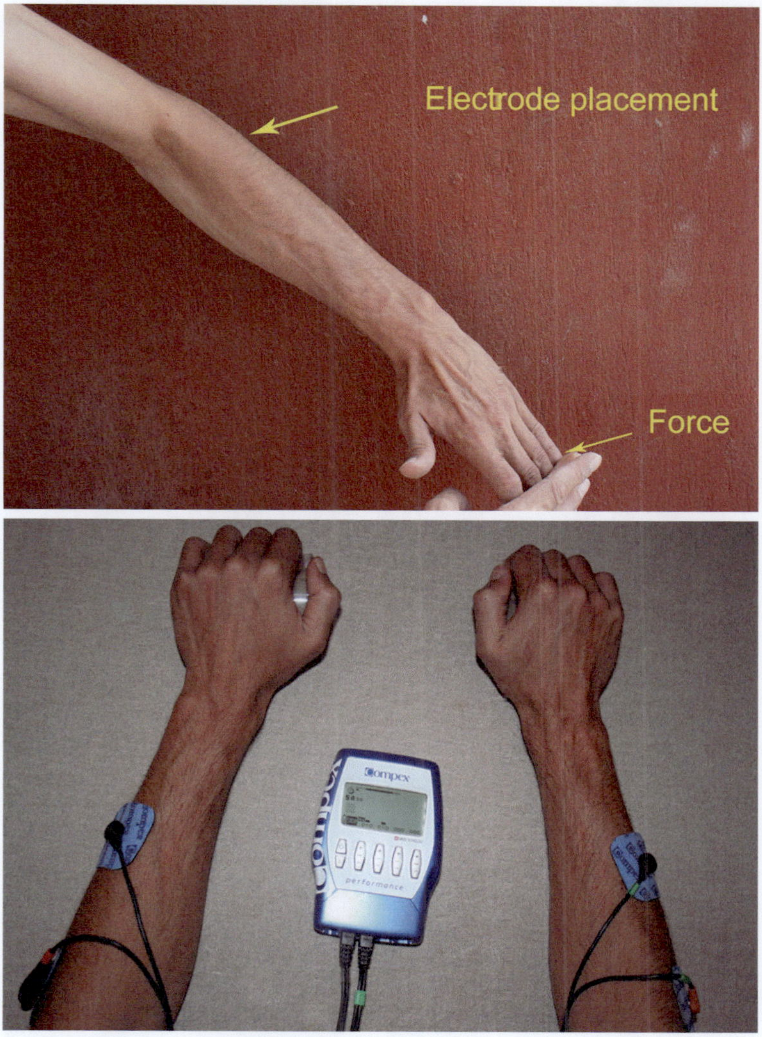

Photos 6.40 and 6.41

This operation can be done to set the electrodes on the flexor muscle.

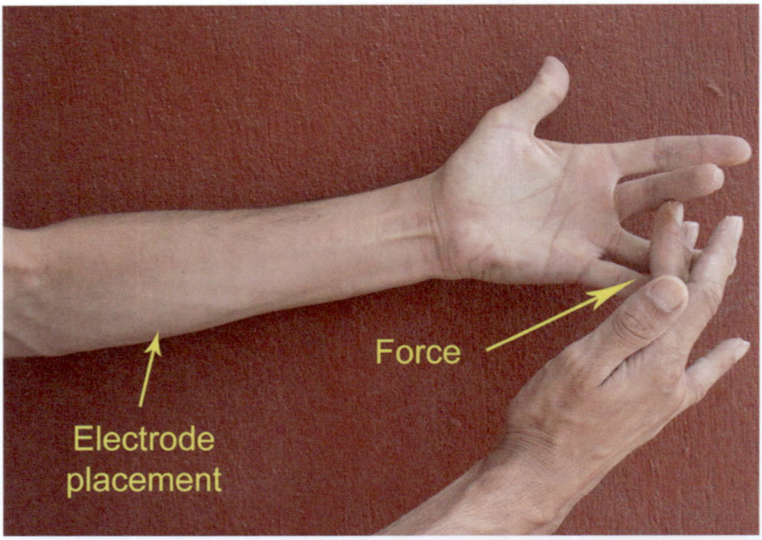

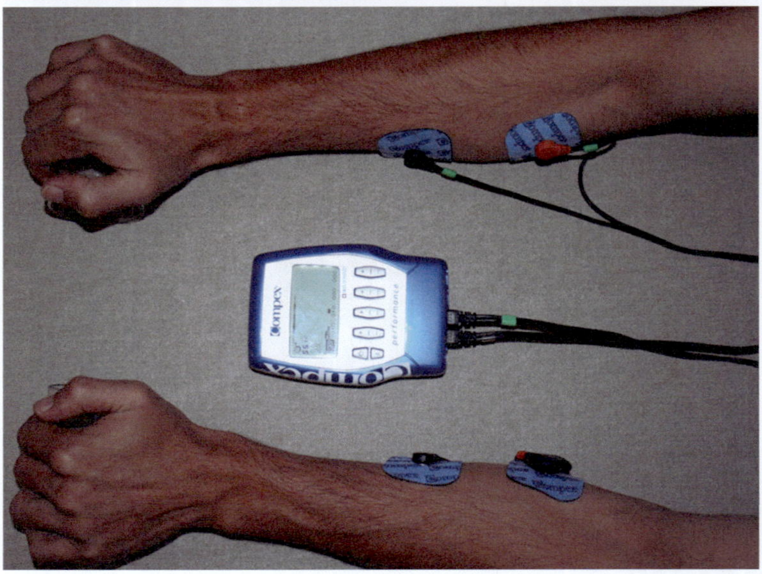

Photos 6.42 and 6.43

6.7.2.2 Analgesic Effects

Throughout that exercise, the finger must not be contracted. Pins and needles can be felt, but the important thing is to feel a certain relief even if it is not immediate.

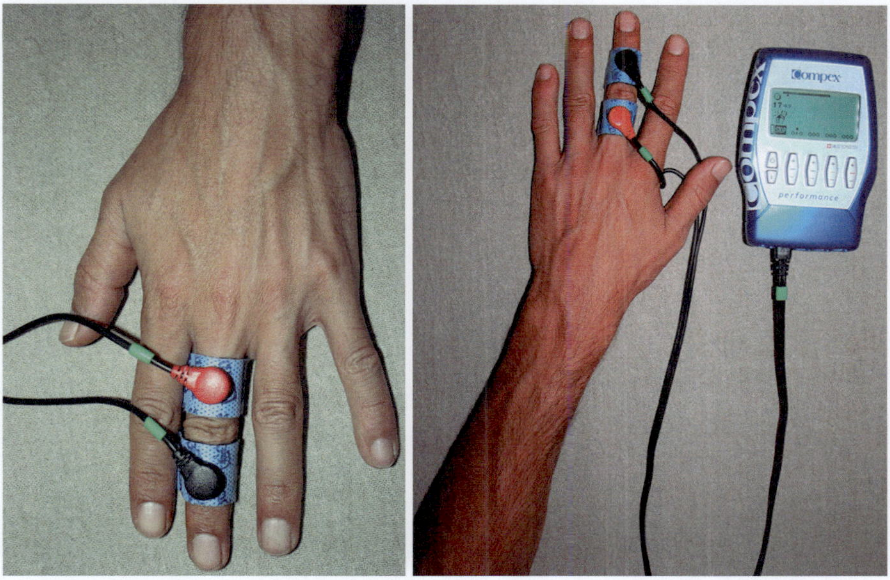

Photos 6.44 and 6.45

Photo 6.46

6.8 Osteoarthritis

Because of the specificities of climbing and as climbers get older, pains in the PIP and DIP joints can be felt. This soreness can be compared to osteoarthritis which is a failure in joint cartilage. But is it really osteoarthritis and can it be said that climbing is a factor triggering this problem?

What is osteoarthritis? It is a painful disease due to an early wearing out of joint surfaces. Finger movements are usually helped by the smooth surface of the bones also known as the cartilage. Cartilage may happen to wear out before cracking. Once the original smooth surface is damaged, a bone excrescence appears and prevents the normal sliding movement. Shocks are no longer stopped and bones keep degenerating which lead to ultimate pain. The soreness changes according to the movement but also to the load undergone by the finger. Sometimes after a rest period, fingers are difficult to put into action. Joints can get stiffer and movements harder to make.

An intensive use of the joint due to a biomechanical pressure and an excessive fatigue could be the cause of this problem. In climbing, the recurrent fast movements and physical efforts in extreme positions may damage the cartilage and consequently the joints. Nonetheless, the origin of osteoarthritis is still discussed by several reports. Osteoarthritis can be detected by X-rays.

The symptoms and the causes of osteoarthritis are felt throughout the climber's life. It is quite common to see climbers complain about finger pains as time goes by and finally realize that they are touched by osteoarthritis. But can it be said that climbing practice is responsible of this phenomenon? Can climbing practice be a factor in the appearance of osteoarthritis?

Researches seem to claim the opposite. A late study made by Sylvester, Christensen, and Kramer has come to the conclusion that climbers are not more concerned by osteoarthritis than ordinary people. Of course fingers are constantly put under pressure especially when they have to support the load of the entire body in specific movements. The more advanced the climber is, the more important is the strain involved on fingers. Obviously due to the intensity of the practice, the body is supposed to strengthen in specific parts in order to keep the right balance between the various loads and tensions. This strengthening of the body affects the climbers' finger bones and is noticed by the broadening of the PIP joints to put up with the biomechanical stress. Even when climbers start practicing as they are grown-ups, the body will naturally react by developing the size of the bone in the metacarpals and the phalanges.

It appears that climbers' bones are tougher than ordinary people's. Moreover and according to their fields – bouldering, routes, etc. – the more they practice, the tougher their bones get. Bouldering practice brings a higher pressure on finger joints due to extreme movements on smaller holds. Therefore, this practice leads to a more important thickening of the bone.

According to Schoff and Al, the impact of finger mechanical pressure can't be properly defined in the long term. In youngsters, the consequences of this phenomenon are a cortical hypertrophy, a PIP broadening, and a subchondral sclerosis. Osteoarthritis cases are quite rare. X-rays have been done on 65 climbers by Rohrbrough and Al, and they did not manage to prove for sure that osteoarthritis affected more climbers rather than ordinary people.

It is quite usual to see youngsters in high-level competitions whether in bouldering or lead categories; this is why searches will have to be carried out to confirm the previous results.

Photo 6.47

6.9 Icing Treatment

Icing treatment is often recommended in case of injuries. This habit and the effects of icing on the limbs are well known by all sportsmen. Icing is a good anti-inflammatory and healing treatment. Different forms of icing such as sprays or ice cube bags can be used in an efficient way.

Icing is a good means to reduce a bruise or an edema. Blood circulation is also restored to a slower flow by vasoconstriction. A lowering in body waste and toxins is also noticed. Besides, it is a good pain killer.

6.9.1 Icing Bag

To be efficient, ice has to be applied every 7 min for about 10–15 min. The application must not exceed 20 min, and the skin must not be in contact with the ice. Ice cubes have to be wrapped up in a linen. As for the ice bag, it has to be put into a wet towel.

Another technique consists of applying ice for 10 min on and to wait for 20 min to let the blood flow in the injured part before icing again for 10 more minutes.

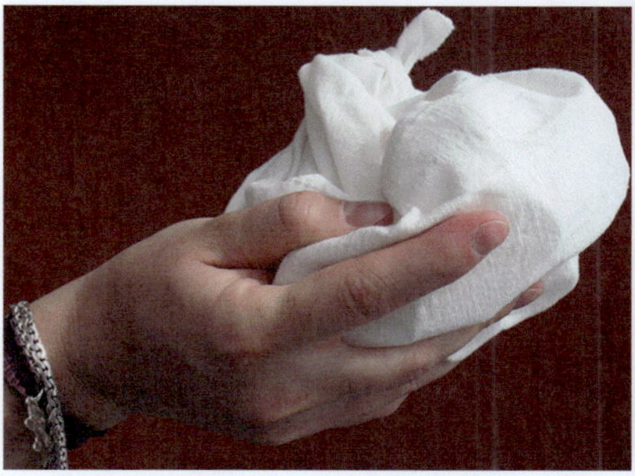

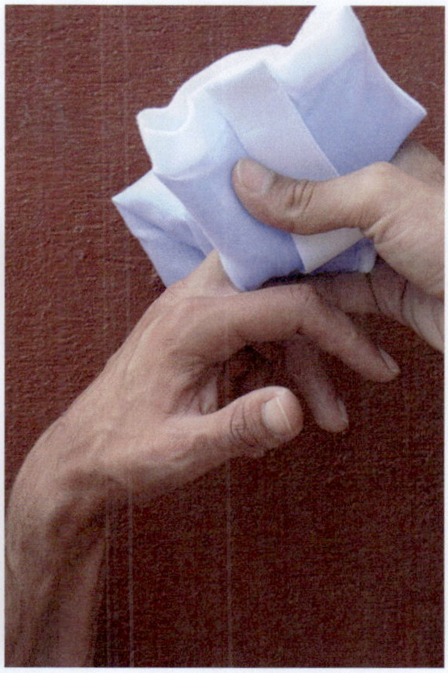

Photos 6.48 and 6.49

6.9.2 Partial Immersion

Icing is not only a good means to kill pain or reduce swelling but also a way to ease recovery. After an effort, the technique consists of immersing the overused part of the body in a mixture of ice and water at a temperature between 2° and 10°. The time of immersion depends on the climber's limits to the cold.

In climbing, it is possible to immerse the hands and forearms which allows a better adjustment to the amount of training sessions and helps recovery. Therefore, it is advised to follow these recommendations especially during intensive and long training sessions which always are a real strain on finger structures.

During an intensive cycle, the climber can immerse some parts of his body for 10 min, putting the limb concerned in and out the water every 30 s.

Photo 6.50

A few precautions have to be taken. *It is strictly forbidden to apply ice directly on the skin.* Cold benumbing effects reduce gesture precision. It is consequently extremely dangerous to train right away after ice application. *Applying ice before any effort is to be banned. Icing must never be applied on an open injury.*

In summary, icing can treat pain, inflammation, and blood circulation problems and help promote recovery and relaxation.

Photo 6.51

6.10 RICE Method

The RICE method – rest, ice, compression, and elevation – is the basic principle to apply in the case of minor injury. The sooner the athlete reacts, the better the recovery will be. Indeed, this method is a good means to relieve swellings, sprains, or bruises.

6.10.1 Rest

A sore sensation never has to be overlooked because it is the body's way of saying "stop." It is a sign which has to be taken into account. It is the sign which tells the climber that the injured limb needs some rest. In that case, no weight has to be supported by the finger, and a complete and immediate interruption of practice is recommended.

6.10.2 Ice

The first thing to do is to apply ice on the injured limb. It will reduce the swelling. More details about icing effects have already been given in the previous part dealing with icing treatment.

6.10.3 Compression

After icing the injured part of the body, it is recommended to bandage the area to reduce the swelling too. It is possible to tape the lesion, but not too tight to allow blood circulation. The climber's fingertip must neither get colder nor change color. The finger has to be maintained in a straight position.

6.10.4 Elevation

The injured limb has to be elevated above the heart level. The injured hand can be elevated by means of cushions which will help reduce the swelling, thanks to the action of various hydrostatic pressures. Raising the wounded hand also eases blood circulation toward the heart.

After a couple of days, many minor injuries start to heal up. After 48 h, if the swelling is still visible, it is advised to consult a specialist to make the proper diagnosis and prescribe the adequate treatment.

Photo 6.52

Chapter 7
Conclusion

Throughout this book, we have dealt with the specificities of hand injuries resulting from that fascinating though demanding sport that climbing is. Fingers are the key part to achieve a great performance. Most of the time, climbers account for their failure by mentioning a lack of strength in their fingers, putting aside all the other factors which may have been involved in that bad performance. Consequently, climbers tend to overuse these fingers, doing training sessions more and more intensive and sometimes up to the limits. Therefore, it is a priority to bear in mind all the factors involved to succeed but also to strengthen all the proper muscles in order to reduce finger injuries.

To know better the body functioning helps reduce traumatizing injuries. Some *vital* rules have to be respected whatever sport is practiced. The body has to be listened to because it is far from being a machine. This is why any sign of pain has to be taken into account and regarded as a sign of *dysfunction*.

Warm-up, rest, progression, and stretching exercises are factors to be considered to last in climbing practice. Great results –routes 9a+– are achieved by youngsters aged 15. Only the future will tell us what will happen to these young athletes especially as regards their hands' state.

Healthy fingers could become a key word to maintain a long career at high level. Just like all the other sports activities, the most difficult thing to do is to maintain a high level and not to achieve it. Finger injuries leading to a certain interruption of practice can prevent this constant level. The climber has to be aware of that and act safely to limit any lesion risks.

Nonetheless, even if safety and caution are respected, some accident may happen. Handholds, movements, and the type of movement can result in dysfunction of the finger structures. Anyway, the first treatment is of course rest and then the complete cessation of practice before consulting a specialist to know the proper diagnosis and treatment. Later on, the coach and the climber will have to find the causes of the injury to avoid any recurrences.

S. Gnecchi, F. Moutet, *Hand and Finger Injuries in Rock Climbers*,
Sports and Traumatology, DOI 10.1007/978-3-319-16790-9_7

To conclude, to recover one's former level, it is better to get back on track slowly and smoothly. The most difficult thing to do is to convince the climber to be patient, but this progressive work is necessary to assure his future in climbing in the long term. If he happened to disobey this basic principle, his sports career might be shorter than what he expected.

Photo 7.1

Abbreviations

A1 A2 A3 A4 A5	Arciform pulleys
ATP	AdenosineTriphosphate
BMI	Body Mass Index
C1 C2 C3	Cruciform pulleys
CDLL	Carpal Dorsal Littlefinger Ligament
CG	Climbing Gym
DIP	Distal Interphalangeal joint
DTM	Deep Transverse Massage
EDC	Extensor Digitorum Communis
ES	Electrical stimulation
ESS	Electro-Skin Stimulation
FDP	Flexor Digitorum Profundus
FDS	Flexor Digitorum Superficialis
GI	Glycemy Index
M	Metacarpal
MCP	Metacarpal Phalangeal joint
MRI	Magnetic Resonance Imaging
NSAID	Nonsteroidal Anti-Inflammatory Drug
P1 P2 P3	Phalanges
PC	Phosphate Creatine
PIP	Proximal Interphalangeal joint
PP	Physical Preparation
RCL	Radial Collateral Ligament
RICE	Rest Ice Compression Elevation
SP	Specific Preparation
UCL	Ulnar Collateral Ligament

S. Gnecchi, F. Moutet, *Hand and Finger Injuries in Rock Climbers*,
Sports and Traumatology, DOI 10.1007/978-3-319-16790-9

Climbing Lexicon

On sight to succeed a passage on the first attempt without any preparation.
After work to succeed a passage after several attempts.
A bucket a large hold.
A boulder a surface on which the climber is in action without any belayer.
Hanging to maintain a position during a hangboard exercise or to reach another handhold by helping oneself with one hand.
Ranking classification of difficulties for a route.
Difficulties route competitions
A dynamic movement a fast movement.
A dyno a fast movement to reach a faraway handhold. Feet or hands are no longer in contact with the surface.
No foot a movement without any foothold.
A campus board a wood device to work on specific movements.
A sloper a round handhold.
A small edge a small handhold.

S. Gnecchi, F. Moutet, *Hand and Finger Injuries in Rock Climbers*,
Sports and Traumatology, DOI 10.1007/978-3-319-16790-9

Bibliography

Bernhardt M, Banzer W, Weipert H (1999) Injuries at interphalangeal joints in freeclimbers and its related risk factors. Dtsch Z Sportmed 50:78–82

Bonola A, Caroli A, Celli L (1998) La Main. Piccin, Padova

Booth J, Marino F, Hill C, Gwinn T (1999) Energy cost of sport rock climbing in elite performers. J Sports Med 33:14–8

Chevrot A (1998) Imagerie clinique du poignet et de la main. Masson, Paris

Cometti G. Intérêt de l'électrostimulation dans l'entraînement des sportifs de haut niveau, Université de Bourgogne

Danowski RG, Chanussot JC (2005) Traumatologie du sport. Masson, Paris

Doyle JR (1988) Anatomy of the flexor tendon sheath and pulley system. Journal of Hand Surgery 13:476–84

Dufour M (2002) Anatomie de l'appareil locomoteur : membre supérieur. Masson, Paris

Dupuy C (1991) Escalade ; acte du colloque ENSA Chamonix 89. Actio, Joinville-le-Pont

Duval MA (1986) La Main du grimpeur : approche physiologique, clinique et expérimentale, thèse de doctorat de médecine, Nice

Garnier A, Waysfeld B (1992) Alimentation et sport. Maloine, Paris

Grégoire R, Oberlin S (1973) Précis d'anatomie. In: Tome I (ed) Atlas, 9th edn. Baillière, Paris

Guyon L, Broussouloux O (2004) Escalade et Performance. Éditions Amphora, Paris

Kahle W, Leonhardt H, Platzer W (1996) Anatomie, appareil locomoteur. Flammarion Medecine-Sciences, Paris

Klauser A et al (2002) Finger Injuries in Extreme Rock Climbers Depiction with Dynamic US. Radiology 222:755–61

Le Bourg M (2006) Les Traumatismes fermés récents des IPP des doigts. Rev Chir Orthop 92: 1S83–1S8

Leclerq C (2001) La Main traumatique du sportif. Masson, Paris

Lestienne F, Feldman A (2002) Une approche théorique de la production du mouvement: du modèle lambda au concept de "configuration de référence productrice d'actions", Sciences & Motricité, n° 45, 9–43

Merle M, Dautel G (1992) La Main traumatique, 2nd edn. Masson, Issy-les-Moulineaux

Middleton P, Trouve P, Puig P (1994) Étude critique des rapports agonistes/ antagonistes concentriques chez le sportif. In: Simon L (ed) Actualités en Rééducation Fonctionnelle et Réadaptation. Masson, Paris, pp 18–22

Moutet F (2003) Flexor tendon pulley system : anatomy, pathology, treatment. Chirurgie de la main 22:1–12

Moutet F, Guinard D, Gerard P, Mugnier C (1993) Les ruptures sous-cutanées des poulies des fléchisseurs des doigts longs chez les grimpeurs de haut niveau. Ann Hand Surg 12:182–8

S. Gnecchi, F. Moutet, *Hand and Finger Injuries in Rock Climbers*,
Sports and Traumatology, DOI 10.1007/978-3-319-16790-9

Pageaux C, Magnan D, Riché D (1990) Une nouvelle chance pour le sport. Les apports nutritionnels. Sport et Vie, Dijon

Paillard T, Noé F, Edeline O (2005) Effets neuromusculaires de l'électrostimulation transcutanée surimposée et combinée à l'activité volontaire : une revue. Annales de réadaptation et de médecine physique 48:126–37

Pradet M (1999) La Préparation physique, coll. « Entraînement », INSEP Publications

Quaine F, Vigouroux L (2004) Maximal resultant four fingertip force and fatigue of the extrinsic muscles of the hand in different sport climbing finger grips. International Journal of Sport Medicine 25:634–7

Quaine F, Vigouroux L, Martin L (2003a) Effect of simulated rock climbing finger postures on force sharing among the fingers. Clinical Biomechanics 18:78–84

Quaine F, Vigouroux L, Martin L (2003b) Finger flexors fatigue in trained rockclimbers and untrained sedentary subjects. International Journal of Sport Medicine 24:424–7

Rohrbough JT, Mudge MK, Schilling RC et al (1998) Radiographic osteoarthritis in the hands of rock climbers. Am J Orthop 27:734–8

Roloff I, Schöffl VR, Vigouroux L, Quaine F (2006) Biomechanical model for the determination of the forces acting on the pulley system. Journal of Biomechanics 39:915–23

Salomon JC, Vigier C (1989) Pratique de l'escalade. Vigot, Paris

Schöffl V, Hochholzer T, Imhoff A (2004a) Radiographic changes in the Hands and Fingers of Young, High Level Climbers, Am J. Sports Med 32:1688–94

Schöffl V, Klee S, Strecker W (2004b) Evaluation of physiological standart pressures of the forearm flexor muscles during sport specific ergonometry in sport climbers. J Sports Med 38:422–5

Schweizer A (2000) Biomechanical effectiveness of taping the A2 pulley in rock climbers. Journal of Hand Surgery 25B:102–7

Schweizer A (2001) Biomechanical properties of the crimp grip position in rock climbers. Journal of Biomechanics 34:217–23

Schweizer A (2003) Lumbrical Tears in Rock Climbers. The Journal of Hand Surgery 28:187–9

Schweizer A, Frank O, Ochsner PE, Jacob HAC (2003) Friction between human finger flexor tendons and pulleys at high loads. Journal of Biomechanics 36:63–71

Shahram A, Farzad A, Reza R (2007) A study on the prevalence of muscularskeleton injuries of rock climbers. Physical Education and Sport 5:1–7

Simon LS (1999) Osteoarthritis : a review. Clin Cornerstone 2:26–37

Sylvester AD, Christensen AM, Kramer PA (2006) Factors influencingosteological changes in the hands and fingers of rock climbers. Journal of Anatomy 209:597–609

Tubiana R (1980) Traité de chirurgie de la main. Masson, Paris

Vigouroux L, Quaine L (2006) Fingertip force and electromyography of finger flexor muscles during a prolonged intermittent exercise in elite climbers and sedentary individuals. Journal of Sports Sciences 24:181–186

Vigouroux L, Quaine F, Labarre-Vila A, Moutet F (2006) Estimation of finger muscle tendon tensions and pulley forces during specific sport climbing grip techniques. Journal of Biomechanics 39:2583–92

Vouillaume D, Forli A, Parzy O, Moutet F (2004) Réparation des ruptures de poulie chez le grimpeur. Chirurgie de la main 23:243–8

Weineck J (1983) Manuel d'entraînement. Vigot, Paris

MIX
Papier aus verantwortungsvollen Quellen
Paper from responsible sources
FSC® C105338

If you have any concerns about our products,
you can contact us on
ProductSafety@springernature.com

In case Publisher is established outside the EU,
the EU authorized representative is:
Springer Nature Customer Service Center GmbH
Europaplatz 3, 69115 Heidelberg, Germany

Printed by Libri Plureos GmbH
in Hamburg, Germany